PEOPLES OF THE FLUTE

A study in anthropolithic forensics

Bob Patten

STONE DAGGER PUBLICATIONS
Denver, Colorado

This book is intended to educate and entertain. The author shall have neither liability nor responsibility to any person or entity with respect to any loss or damage, directly, or indirectly, by the information contained in this book.

Cover art by Bruce Burks. Illustrations on pages 247, 248, 253, and 254 were drawn by Lauren Patten. Images on page 107 courtesy of Peter Kelterborn. Unless otherwise attributed, remaining illustrations are by Bob Patten.

 For information contact *Stone Dagger Publications* 10803 W. Connecticut Ave. Lakewood, CO 80232.

StoneDagger.com

Library of Congress Control Number: 2005901062

Patten, Bob

Peoples of the Flute: A study in anthropolithic forensics.

Patten, Bob, 1944-

ISBN 0-9668701-1-5

Cover

The hump-backed Kokopelli first appeared in rock art as a trader with a back pack. With that in mind, he becomes more than a visual pun—he represents an emissary spreading new ideas in lithic technology.

Cover painting in egg tempera by Bruce Burks

Peoples of the Flute:
A study in anthropolithic forensics

Table of Contents

To the late George Stewart, trapper, flintknapper, and friend, who recognized that the best application of lithic technology was sharing it.

Acknowlegements

Countless writers have been responsible for stirring my imagination about how fluting began. The questions and ambiguities each author presented were as important as the facts. How could so much data add up to so much dissention? Now that I am ready to add my own contribution, I am forced to recognize that my perspective may be as biased as any before me. So, my thanks are extended to everyone who has ventured to speculate on the how, why, where, and who of the earliest Americans. Each insight builds toward a richer understanding.

Only in retrospect did I truly appreciate the collaborative role my wife Laurey played in shaping the emerging story and encouraging new insights. Her exhortations to "dig deeper" and "be a better tour guide" allowed me to capture nuances intended, but not yet effectively conveyed. By coining the term "anthropolithic," she improved the direction and focus of the book. Laurey's artistic talent is seen in some of the sketches of knappers in action. I owe much to her patience with my long-standing obsession, and to her constant encouragement to realize my dreams.

The detailed discussion of fracture is due to the tireless efforts of Tony Baker, who grew up in a family dedicated to archaeology, but detoured to a career in petroleum engineering. Upon retirement he discovered a novel way to apply his professional talents to archaeology via computing the behavior of fracture. Our frequent and fruitful conversations have generated many new insights. In addition, Peter Kelterborn and Andrew Pelcin laid important groundwork with controlled experiments in fracture, while Jacques Pelegrin added a valued European perspective. Bill Watts developed mathematical models that clarify impact mechanics and flake paths.

Dennis Stanford and Pegi Jodry of the Smithsonian Institution have provided years of friendship, and access to important archeological collections. They also share an enduring interest in the people who practiced fluting. John Clark, of B.Y.U., has been an significant source of support and was instrumental in assembling experts to participate in two Folsom workshops in 1997 and 1998. Those experts have collectively provided many of the insights necessary for laying out a broad outline for the fluting tradition.

My enduring thanks to each of those who provided support in bring this book to press.

Chapter 1

Introduction

This is a study of those early occupants of North America who were in the habit of knocking large flakes from the bases of their projectile points, a feature archaeologists call *fluting*. Fluted projectile points have a special aura about them because they are so distinctive and unusual. Clovis projectile points, with their relatively short basal thinning flakes, are the first in a series of styles that culminate with Folsom, where the faces are nearly completely removed. Modern attempts to duplicate the style have a notoriously poor record of success when it comes to removing so much of the face of the projectile. Why would people make their stone spear points in such an apparently risky way when the points were so important for survival? Where did the practice come from, since no direct antecedents have been located? Why did fluting practice suddenly cease after being so popular and widespread? Generations of archaeologists have pondered these questions with relatively little success. Experiments in replicating fluting technology at times suggest possible answers but invariably raise as many questions as they appear to solve. There are so many methods for making fluted points that it sometimes seems no single solution can ever be agreed upon. Could leverage, punch, pressure, and percussion all have been used in turn, depending on the knapper's whim? Since its recognition in the 1830's (Tankersly 2003), the why, how, and origins of fluting have defied consensus. Yet a surprisingly simple solution can be advanced. Ultimately, the key question is whether early people invented and used new techniques at will, or if one discovery evolved naturally to the next.

Forensic anthropolithics can be used to solve the mystery. Since the term *anthropo-lithic* means "made out of stone by the hand of man," it highlights the human influence on stone artifacts. Forensic is defined as "suitable for public debate," so one ought to be able to see and question the clues that lead to a given conclusion. In fact, an audit trail should be developed for each line of evidence, to show how that line relates to the whole. Science need not be hidden, or cryptic, or even purely objective. Dispassionate consideration of processes and their results has its place, but we are talking about people. By including motivations and responses, the deeper meaning in ancient artifacts can be brought to light. After all, most of us are interested in archaeology because we are curious about our ancestors. It's not just where we come from, but how we spread, and why we make the choices we do. Yesterday doesn't have to be far removed from today if the right perspective is applied.

Human behavior is as much a part of nature as that of butterflies or any other creatures. Recognizing the relationship between human behavior and the laws of nature helps identify principles that control the advance of lithic technology. While this book focuses on how and why humans developed fluted points, the same considerations operate with any innovation for any group of people at any period in time. I use fluting technology as a case study, with the expectation that the pattern of human response will prove useful for a wider scope of archaeology.

Any reporter can reel off a series of questions essential to developing a story. In prehistory, *"Who?"* is related to culture and genetics. *"What?"* is basic archaeological data. *"Where?"*

can be answered by site locations. *"When?"* calls for measuring how particular elements change over time. *"How?"* may be extrapolated by observation and trial. *"Why?"* though, is often considered pure speculation and not suitable grist for the scientific mill. Consequently, the question is frequently sidestepped, even though it may really be the most fundamental question of all for a historian. I intend to show as much as I can about why decisions about lithic technology were made. So, as you read this book, don't let the facts take you too far from considering people's motivation. Our forebear's reasons for what they did may not be evident today, but we can be sure they were justified for the time and the place.

In 1997, a workshop was held in Austin, Texas (Clark and Collins 2002) to resolve the mystery of how Folsom projectile points were fluted. As part of his concluding remarks, workshop organizer John Clark of B.Y.U. noted that, while there were many technological routes to the final product of a fluted projectile point, none could be unequivocally identified as *the* unique method used by early practitioners over 10,000 years ago. For some scientists, that conclusion fits comfortably with their notion that such cultures were inherently adaptive. Others argue that early cultures were too bound by tradition to have used multiple fluting approaches at the same time. Evidence detailed in this book demonstrates that all fluting methods used by early Americans arose from a few basic concepts that led to unique results as knapping practice gradually evolved.

Although there are other contenders for the title of oldest artifact style in the Americas, fluting remains the earliest widely accepted type. The simple elegance of a finely-crafted fluted projectile point attracts so much attention that other classes of chipped stone artifacts seem mundane by comparison. Not only does fluting appear as one of the earliest stoneworking traditions in the Americas, it aspired to higher standards than most artifact styles that follow. The early technical brilliance that bloomed from an as yet unrecognized source is only rarely approached elsewhere in the world. The apparent decline in knapping ability following fluting traditions is one of archaeology's major mysteries.

History chronicles a progression of life styles through time, but the pattern of that progression is not always clear. Ingrained social organization defines a cultural framework so well that we can look at remains from thousands of years ago and think of it as a *tradition*—a consistent pattern of doing things, clearly passed down through generations, but still changing through time. That means the impetus for change tends to be held in check by tradition, which works to keep things the same. The really interesting chapters of history occur when great change is forced upon people. We will see how the fluting tradition was forged during a particularly turbulent episode of pre-history. People in later times saw relatively benign impetus for change, but still continued to adapt as nature kept up the pressure. The means by which innovations were spread is crucial to how societies were shaped.

This is not a book to teach flintknapping. *Old Tools—New Eyes* (Patten 1999) was written to cover that ground. The goal is to show *why* people acted as they did and demonstrate how clues can be deciphered to tell a comprehensive story. However, the details of what causes fracture are among the most important facts in a lithic forensic investigation. The catch is that the science of fracture hasn't yet been rigorously applied to analyzing stone tools. Until now.

Information provided about physics and the mechanics of fracture relies heavily on the work of Tony Baker (1998), who managed to use computers to model experiments in which flakes were made by dropping steel balls on the edges of glass plates. Additional experiments isolated the crucial variables and proved that the computer model does indeed represent real-world conditions. Broadening the computer models served to explain even more aspects of fracture. At the same time, modeling led us to test previously untried knapping techniques to validate the predictions. Integrating theoretical models with actual results provided a robust and completely representative solution. It also means that the mechanics of fracture can be compared against known methods for fluting and accounts for differences in results.

Since physics may not be everyone's cup of tea, I have kept the explanations simple. So much rests on understanding how human behavior adapts to the inviolate laws of nature that the case must be made well enough to support a forensic argument. I have left complex equations in the computer, but there is enough information provided for the technically inclined student to follow.

It is one thing to understand the theory of lithic fracture and entirely another to wrest the details from the archaeological record. Speculation has its place, but a combination of theory and experimentation allows clues to point with confidence to the manner and method used to create the litter of stone flakes left at an archaeological site. It is important to avoid the personal bias that purely experimental flintknapping can impart. When one is unaware of another solution, knappers can be tempted to peddle what they have as the best thing yet. Objective forensics makes sure that the site evidence and experimental data each pass the same tests, and combine to make the best case possible.

Archaeological data has often been recorded without a clear theoretical foundation. New measurements, reflecting a fresh understanding of what is relevant, can lead to interesting insights. Artifact collections that have detailed records of their location and context are a most valuable resource for this purpose and allow immediate feedback. This book utilizes mostly summary charts, but due to the diligence of countless investigators, a treasure trove of detailed data in field notes and site reports is available elsewhere for reference.

A paleo craftsman making a new spearpoint turned out a product by using process steps and quality controls, in direct parallel with a modern assembly line. That parallel helps us understand the making of stone tools in familiar terms. It also leads to a life-cycle view of stone implements. Flintknapping demonstrations often mislead people into thinking that artifacts were typically built in a single session. I will try to dispel that notion by clarifying how each stage of work supported a different aspect of daily life. In the process, we will see how constant cycling of work stages both leaves valuable archaeological clues and helps our understanding of them.

One might expect that once people knew how to get stone to break predictably, they could make whatever style stone tool they chose. However, it is remarkable how one can produce a good flake and then find the next one nearly impossible to accomplish. It's rather like a 3-D chess game in which it is necessary to anticipate how each move sets up possibilities for all the following moves. Strategies for taking rocks apart are boundless but, by following certain patterns, a craftsman can keep variation in check. Interpretation of stone tools is significantly facilitated when patterns of flake scars can be seen as a means of satisfying a functional end rather than simply as aesthetic design. The way decisions are made reveals what is truly important.

Without the efforts of artifact replicators, I would be unable to say with any assurance how projectile points were fluted. Like wending one's way through a maze, until we know the full layout there is no choice but to investigate where each alley leads. For this reason, previous theories, with all their flaws and gaps, show the way as much by their failings as by their success.

Scientific resolution is an ongoing cycle. Rather than being at the punch line of the story, you are about to embark on an unending voyage of archaeological discovery. Strong evidence allows us to reconstruct the inception, development, and abandonment of fluting. There are, however, gaps in the record that remain to be explored. The outline presented here is meant to help archaeologists structure their efforts to fill the gaps. Many of the concepts are new and need to be tested and refined by applying them to a variety of problems. Computer modeling of fracture is very much in its infancy and will undoubtedly benefit from the rapid rise in computer applications. That kind of knowledge will inevitably allow reconstructing a knapper's physical action. Each advance in understanding fracture mechanics also provides a more secure basis for deciding which parameters should be measured, and for understanding what the data represents.

Just as one site cannot be used to characterize an entire tradition, no one investigator is able to portray the sweep of ancient life without leaning on the work of others. The task is an unending repetition of playing new data and insights against previous ones, and any conclusions must be considered temporary. We should expect new data, new concepts, and new analytical tools to move the story of archaeology dynamically.

Chapter 2

Evidence

As in any other forensic endeavor, a first step in tracing the origins and significance of fluting technology is to gather the relevant evidence. In the early stages, the pile of evidence can be intimidating and confusing. Throwing out inconvenient data is particularly unwise until we know why they do not fit—they may actually be the best clues we have. With that in mind, this book will consider a melange of evidence to see where the chronicle needs bolstering and what questions are important. At this point, it is important not to jump to conclusions, and simply treat the information as raw data. Somewhat like starting to solve an old jigsaw puzzle, we need to assess what may be missing and what may be extraneous. Data that does not fit shows where evidence needs to be bolstered.

Archaeologists decide where to dig based on their experience from previous encounters with similar material. They are motivated to maximize what they can learn with limited resources and cannot afford to look in random places, so anomalies can go undiscovered. Investigators at each site reviewed in this chapter were obligated to provide interpretations of the data they collected, leaving us with a sometimes-confusing, often contradictory set of assertions. So, a number of questions regarding fluting technology remain open: *How did fluting practice originate? Did Clovis ancestors arrive from the east or from the west? When did they arrive? How could fluting have been dispersed so rapidly over such a wide area? What caused the practice of fluting to disappear?*

Every investigator comes at the problem from a different background with built-in biases and perspectives. This is inevitable because, although new data is constantly being added, the archaeological record will always be incomplete. Such gaps force researchers to speculate about the connections, and the nature of academia makes it difficult for new constructs be accepted quickly. Archaeologists are known for caution in adopting new paradigms. As the saying goes, "Once bitten, twice shy," and archaeologists have been bitten more than once by poorly founded assumptions. While conflicting theories abound, it is healthy to continually advance fresh views of old problems, as long as the reasoning is sound. The constant rehashing serves to wring weaknesses out of the interpretations. Most of the persistent ambiguities in the archaeological record have less to do with lithic technology than with dating, genetics, and culture, but no manifestation of human activity can remain independent of another. This chapter is intended as a backdrop of what is generally agreed upon from archaeology. Later chapters will attempt to integrate that information and develop new insights that tell a comprehensive story about fluting technology.

Precursor technological elements

The quest for origins of fluting technology must begin with possible precursors. Since there are no signs of fluting outside the Americas, we have to look both to Europe and Asia for technological attributes that could develop into fluting. We also must be concerned about important technological characteristics that failed to reach the Americas early in spite of being on a

Dating conventions

Archaeologists use some potentially confusing conventions when it comes to specifying the flow of time. Writing about years gone by raises the question of where do you count from? By convention, "present" has been defined as 1950, so 2000 years B.P. (Before Present) is always at one point in time equivalent to 50 B.C. (Before Christ).

Radiocarbon dates reflect the time it takes for carbon isotopes to break down, so if the date was measured by radiocarbon decay it is labeled RCYBP (RadioCarbon Years Before Present). Unfortunately, the rate of isotropic breakdown is not steady and calibrations must be applied to reflect true years elapsed. Radiocarbon dates near 10,000 RCYBP need to have about 2,000 years added to reflect true age. Additionally, radiocarbon dates cease to be useful beyond about 40,000 years ago.

Another complicating factor is that dates are more or less reliable depending on the method used to acquire the date. Serious scholars must return to the primary source of data to resolve discrepancies.

So that the reader can appreciate the proper depth of time, ages have been roughly calibrated.

proposed path of entry. Keep in mind that technology and cultural affiliation are not necessarily equivalent. It is quite possible for a technology to be shared by people that do not even share the same language. People of a common genetic stock may also adopt new technology without changing their cultural identity.

In Europe, stoneworking traditions that utilize principles which might be expected to evolve into fluting manifest themselves as early as 200,000 years ago. Levallois technology removed a predetermined flake by a well-placed blow on a specially prepared core. The resulting unifacial tool was often used without further modification. Mousterian technology evolved about 85,000 years ago from the Levallois approach, increasing the use of flakes made from discoidal cores as the basis for more formal tool forms. Eastward movement of the Mousterian into Asia appears to have been largely characterized by the abandonment of the use of specialized flakes as a basis for tools, and increased dependance on crude bifaces where the remaining core is a tool of itself. Starting about 35,000 years ago, the Upper Paleolithic cultures of Europe are known for their reliance on long, parallel-sided blades from specially prepared cores.

It is necessary to explain the significance of blades. In archaeology, the term *blade* can be used to refer simply to a knife blade, but there is also a very specific definition that reflects a special intent. Blade flakes are defined as being purposefully made over twice as long as they are wide, as part of a series of blade flakes. Organizing flakes in a series successively creates guiding ridges for each new blade, so they can be made with wonderful control and regularity. They can be thin or thick, straight or curved, depending on the purpose they are to serve and the manner of their making. Unless they become the foundation for formal specialized tools, blade flakes are seldom important. Those tools can be scrapers, knives, projectiles, or inserts in another implement.

Flutes are scars left from deliberate use of longitudinally directed slender flakes to thin a projectile for hafting. Because the flake and scar left by a blade is so much like a flute, many analysts are tempted to make a technological connection based on the similarity. Another property that blades share with at least some flutes is that they travel nearly the full length of the core surface. Since flaking techniques were first designed to create functional tools, the relation of cores and bifaces to the blades and flakes removed from them is most readily understood by their ultimate use. Blades are slender unifacial flakes that are useful for the tools made from them. Flutes are useful for symmetrical thinning of a bifacial projectile, while the flake is discarded.

Bifacial technology using edge-to-edge flakes first appeared on the steppes of the Ukraine about 25,000 years ago and spread westward into Europe, where it came to be known as Solutrian. Bifaces are basically cores that are flattened and shaped by flaking on opposing faces. Unlike flake-core technology, where cores are discarded and flakes are kept, bifaces are added to the tool kit. Solutrian technology is restricted in area as well as time. It ended abruptly about 18,000 years ago on the western coast of Spain and was succeeded by the Magdalenian technology, which used blades but made no bifaces at all. Although Solutrian technology did not include fluting, it encompassed so many other attributes in common with American fluting traditions it seems a viable candidate for the technological origins of fluting. Like the New World Clovis people, Solutrian tool manufacturers used hammers made of antler batons to make large, flat bifaces with flake scars of considerable extent. Spectacular examples of those bifaces were cached with ochre coatings similar to the way Clovis caches were treated. Bone and ivory rods identical to those found in Clovis sites extend the cultural parallel. In fact, aside from the absence of flutes and some differences in projectile outline, Solutrian use of blades as formal tools is the main point of departure from Clovis technology. Curiously, shouldered Solutrian points have sharp edges while the other forms are ground along the basal margins, as are American Paleo point styles. Even though Solutrian technology provides viable precursor traits (Chandler 2001), there is still the question of how people could have carried the tradition to the Americas from Europe.

The Atlantic Ocean would appear to present an insurmountable barrier to the Solutrian migration to the Americas, until conditions of the time are taken into account. An ice sheet stretching from the coast of France to the Grand Banks of Newfoundland would have presented a rather direct route, if ocean travel were accepted. Weather would have been buffered by the ice front, and currents would have caused a direct conveyor effect between continents. For archaeologists to accept this source of Clovis technology requires that they account for an apparent lack of physical manifestation for up to 5,000 years after arrival in the Americas before selected aspects of European stoneworking concepts were resurrected. A serious concern shared by many archaeologists about an eastern entry is that it would have occurred too rapidly to easily account for the abandonment of critical technological traits, like reliance on blades as the basis for important tools.

A poorly known variant of Solutrian along the coasts of Portugal and Spain represents a substantial shift from the popular

Prior art

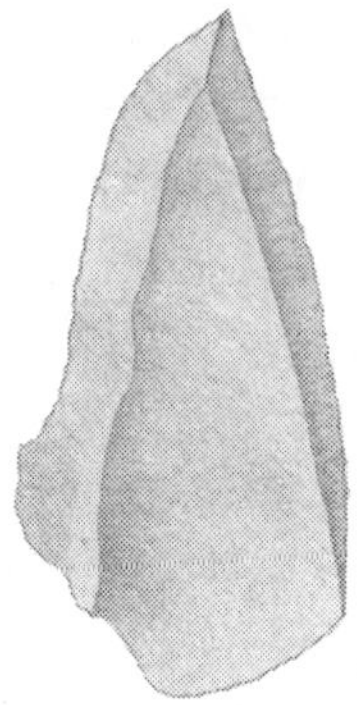

A.

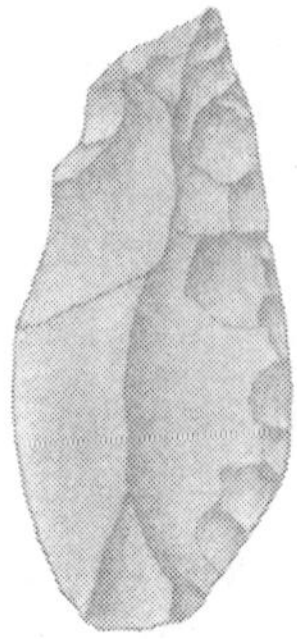

B.

C.

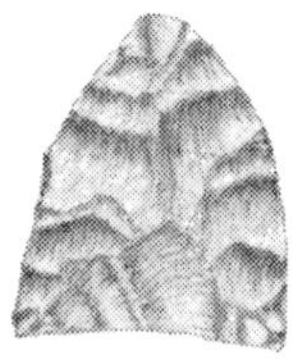

D.

A) Levallois points made good use of the core surface by carefully structuring an impact site that removed a single full-length flake in a predictable location.

B) Mousterian technology used discoidal cores and produced crude bifaces.

C) Solutrian bifacial technology was very controlled, with flake scars extending well past the midline.

D) Point from Cactus Hill, VA. may be nearly as old as Solutrian. Its profile is quite flat.

model of shouldered points and heavy reliance on blades as tools (Chandler 2001). In this manifestation, the points strongly resemble unfluted Clovis points with concave bases. Flake scars carry well past center to make relatively flat, lanceolate projectiles. Shell middens and sea animal remains indicate a people committed to exploiting ocean resources, unlike the big-game hunting we see depicted in popularized accounts. Were Solutrian technology to be exported across the Atlantic, such maritime skills would be essential. That same maritime bias could also mean that important coastal evidence of early migrations was largely destroyed by rising water levels at the end of the Pleistocene.

Keep in mind that the vast majority of Solutrian archaeology was done at a time when much of what was dug up was put in storage, with little or no further analysis. Debitage often has been considered useless, and discarded. Since modern techniques have been in use, most new sites have been funded and investigated to address local questions rather than regional issues. There is need for updated synthesis of the full scope of archaeological data so that future investigations can take advantage of a larger view. Another concern is that the definition of Solutrian culture is out of date, and may not adequately reflect variation caused by differences in local resources. For example, Les Maitreaux in France is the only Solutrian quarry site that has been explored, so far. Refitting of flakes from the site shows that substantial bifacing was done at the quarry, in stark contrast to Clovis habits of reducing rock at a distance from the quarry.

Recently, the Cactus Hill site in Virginia, has revealed evidence that may date over 17,000 years ago. The projectiles have a Clovis outline, but are flat and lack flutes. Some small blades are present, although the prevailing meta-quartzite stone is too grainy to allow clearly distinctive fracture details. Unlike many other early sites, Cactus Hill contains a wide range of camp tools necessary for daily living. Once the site was explored, similar artifacts from unstratified surface locations were recognized as well. This, along with the Topper site (Goodyear 2003), in South Carolina, and other comparable sites along the East Coast, leads some archaeologists to suspect that more pre-Clovis candidates may be found as researchers widen their view of what to look for and where it may occur. Closing the gap between Solutrian times and early American occupations still leaves some critical questions about technological discontinuities between the continents.

Increasing numbers of finds of artifacts in ancient context along the Eastern seaboard must be mentioned, though datable sites are sparse. Most of this promising work is in Florida and near the mouth of the Potomac River. Four issues are important: First, many styles, in addition to Clovis, are associated with Pleistocene fauna. Second, the styles share many attributes with Clovis finds. Third, the artifacts are far more numerous than in other parts of the continent for that time. Finally, they are generally made of local stone. A significant span of occupation and depth of tradition may be inferred from these observations. That fits comfortably with an east coast first entry. Many traits of these early points overlap both with Clovis and Solutrian points, including overshots, concave bases, and flat bodies.

Although colonization of the Americas from Asia has been generally accepted as one of the most likely prospects, at the eastern margin of Asia, mountain ranges and hundreds of miles of ice and snow presented another barrier to human expansion. The most severe episode of glaciation to occur in five million years reached its maximum extent about 18,000 years ago. For nearly 60,000 years, the grassy plains of Beringia, connecting Asia and the Americas, supported various grazing animals like mammoth and bison.

Over thousands of years, repeated opening of the Beringia passage allowed the exchange of many species of animals between continents. Asia acquired the camel, horse, wolf, American fox, and marmot. The Americas received mammoth, bison, black bear, grizzly bear, moose and ultimately, humans. Once people passed the Bering Strait, the new arrivals may have used alternate routes into the continent. Shore hopping would have been difficult because many glaciers intruded into the ocean and safe harbors, called refugia, could be reached only by water. Walking the gap between retreating glacier fronts on the western side of the Rocky Mountains may only have been possible after the earliest colonization had already been established. There is absolutely no doubt that people migrated via many routes over thousands of years, the question is when and where they started the parade. As water accumulated in glaciers, it also lowered the sea level by hundreds of feet. In fact, there were 60 to 90 extra miles of continental shelf exposed at the continental edges, and even more in Florida. Only when the ocean level rose, about 10,000 years ago, did the grassy plain of Beringia give way to tundra that supported little game besides caribou. But the ice-free corridor leading to the south was closed between 22,000 and 12,000 years ago, after the earliest Clovis dates.

Tool assemblages on the Asian side of Beringia remained mostly unchanged for nearly 30,000 years. The Dyuktai tradition in Asia can be characterized as an Upper Paleolithic technology derived from Europe. Bifaces are poorly developed, and while blades are found occasionally, they are not the foundation for any formal tool type. Microblades are another story. They are found in nearly every site, even though they are not always numerous. Occasional bone rods with grooved sides show that microblades were inset to make composite tools. Because microblades include sizes that would be unfeasible as insets in composite tools their purpose is not completely clear. It remains to be explained how a microblade-free technology might be transferred through the Arctic when microblades are ubiquitous. Was fluting developed from the concept of microblades? Complete abandonment of major technological traits is rare, and should not be accepted without a convincing explanation. Did ancestral practitioners of fluting arrive as early as 30,000 or 40,000 years ago, prior to microblades?

A recent site, designated RHS, along the Yana River in northern Siberia, far north of the Arctic Circle, suggests that those early dates are reasonable (Pitulko et al 2004:52). The stonework consists of rudimentary flakes made from crude cores, and lacks blades or blade-cores. Beveled bone rods from the site resemble artifacts from Europe and Pleistocene America. Since the earliest of these rods are not accompanied by stone projectile points themselves, they may actually have been used as projectile points. Most importantly, the site proves that humans were well adapted to the most extreme Arctic conditions and had the means and opportunity to cross between continents as early as 30,000-years ago.

A site at Ushki Lake, on the Kamchatka Peninsula in Siberia (Goebel et al, 2003), was once considered the best candidate for a model of a technology that could be ancestral to Clovis. Recent re-dating, however, shows the materials to be of the same age as Clovis materials. Stemmed points occur at about 13,500 years ago, without the presence of microblades, although microblades are found in overlying layers. There were also bifaces that resembled those found in Clovis sites. Since the artifacts do not predate Clovis, they must be viewed as contemporaneous cousins in the presumed path of travel between continents. One site doesn't sound like much, but it is difficult to find useful sites at that latitude due to the antiquity, the sparse level of early populations, and the difficulty of locating sites that have not been ravaged by harsh arctic weather.

On the eastern side of Beringia, the earliest artifacts are indistinguishable from Dyuktai, but are called the American Paleoarctic Tradition. Earliest dates so far confirm a human presence as early as 14,000 to 15,000 years ago. That it has lasted until roughly 1,500 years ago testifies that this technology is well suited for arctic conditions. A few sites, at which organic materials were well preserved indicate that stone was a relatively minor component of the overall technological base. Bone and ivory rods and speartips were common. In particular, eyed bone needles were very important for enabling people to survive in such a harsh environment. Watertight, sewn clothes and boots are still recognized as the most critical part of arctic gear and are an important prerequisite for making light, hide-skinned watercraft.

Although the American Paleoarctic Tradition appears to have derived from an Asian pattern spanning 30,000 years, other toolmaking traditions shared the same landscape. One of the earliest was the Nenana Complex, which appeared as early as 12,800 years ago. Since the Nenana toolkit lacks microblades, it seems to qualify as a potential ancestor for fluting. However, projectile points are small teardrops and triangles rather than lanceolate bifaces. Especially in the interior of Alaska, there seems to have been an ebb and flow of dominant traditions. The Nenana knapping tradition was clearly displaced by about 12,600 years ago by the Northern Paleo-Indian tradition. The Denali tradition then dominated between 11,500 and 10,500 years ago, after which it disappeared. A point recovered from the Jay Creek site and dating to 11,500 years ago tested positive for mammoth blood (Dixon 1993:86,113), the most recent date known for man and mammoth together.

Another early technology that may be important as a potential antecedent of fluting is typified at the Mesa site (Kunz 2003) on the North Slope of Alaska. When grouped with sites manifesting similar technology, the dates range from as old as 13,000 to 12,000 years ago. However, even the earliest dates are too late to explain continental entry. People who occupied the North Slope had no shortage of flint. Any gravel bar on any river is littered with flint and limestone, and most of the artifacts appear to have been made by using hammerstones in a way that does not allow flakes to reach beyond the center of the biface. No microblades are present in the oldest sites, and the projectile points look similar to the later Agate Basin points of the high plains. The absence of scrapers and gravers is puzzling, but processing caribou may have been easier with bone tools. Published reports refer to a single lithic specimen, fluted from both ends, that is accompanied by microblades and probably was made by the later Dyuktai culture (Kunz 2003:34).

The Mesa site reinforces the idea that dissimilar strategies for living in the far north may have overlapped in time rather than succeeding each other in a neatly ordered progression. In fact, a small percentage of artifacts at the Mesa site show evidence of a technology that uses flat, nearly edge-to-edge flakes (Kunz 2003:29-30). The implication is that some of the earliest debitage has been mixed in a way that is only evident by recognizing the mode of flake propagation. The few preforms that reveal extreme flattening by extensive flakes may be a sign of adaptation to using antler hammers to make edge-to-edge flakes as a way of coping with relatively scarce lithic resources. Conditions south of the Brooks Range would have required a shift in lifestyle since flakable stone is scarce and the mix of game is more diverse.

Isolated finds of fluted Clovis points, usually made of obsidian, constitute what is referred to as the Northern Paleo-Indian Tradition. While there is no datable stratigraphic context, blood residues of mammoth on fluted points demonstrate that the technology was in use while mammoth lived in Alaska. Without datable sites, though, it is not possible to know if Clovis expansion arrived from the south or from the north. The lack of cultural change prior to Clovis times in the Arctic argues that if the roots of fluting technology entered America by that path, it had to have developed from either the Nennana microblade tradition or the Mesa biface tradition.

Because we know that Clovis people were present in the Americas far earlier than any archeological dates established in the Arctic, we must conclude that either the archaeological record in the far north has so far simply missed the first entry of people, or that the entry bypassed the Arctic interior. Indirect assessments from genetics and linguistics have been advanced to estimate that it could have taken 2,000 to 40,000 years for people to colonize the Americas.

Several interesting sites in Nebraska and Colorado (Hall 1995: 7) suggest humans interaction with mammoth as early as 20,000 to 22,000 years ago. Since there are no lithic remains at these sites, they were initially of most interest to paleontologists studying mammoth remains. As pieces of bone emerged with kinds of breakage that could not be explained through natural means, archaeologists saw human activity as the most likely cause. Fresh, "green" bone breaks with a distinctively different pattern than weathered bone. When fragments were pieced together, impact points were clearly demonstrated, showing deliberate spacing and planning. Delicate ribs and vertebrae remained intact within fields of broken long bones with two-inch-thick walls. At the La Sena site in southwestern Nebraska, a vertebra set edgewise in the soft windblown loess soil, close to a concentration of bone fragments, apparently was used as an anvil to expedite the breaking. Lack of stone at several similar sites can be taken as evidence that people were scavenging carcasses rather than actually hunting, and should not imply that there was no lithic technology. However, the possibility that a pre-lithic technology existed in North America should be taken seriously.

The Clovis tool kit inventory

Assessing whether Clovis technology was imported from elsewhere, or if it was invented in North America, requires that we consider a broad array of Clovis tools. Fluted projectile points are well known, but the considerable variability in size suggests that there may have been more on the menu than mammoth. Projectiles are just the penetrating tip of a more complex delivery system that could be anything from thrusting spears to spears thrown with an atlatl. Thrusting spears are relatively stout because they must withstand movement of the animal while the kill is conducted. Spears thrown by the aid of a throwing stick are lighter and usually have detachable tips that can be quickly replaced and re-thrown. Arrows need to be much lighter, and can be lofted from much farther away. The time sequence for introduction of the various forms of weapons delivery is deduced from the projectile tips. By this logic, arrows may have been in use as early as Solutrian times (Straus 1985:504), where small barbed points occur. From history, we know that atlatls were used to fight Spanish colonists, so the overlap between weapons systems is extreme.

Those differences in use are generally recognized by differences in width of the hafting element for later periods of time. Clovis projectiles, strangely enough, encompass the entire range of size. Most points found at kill sites are consistent with the use of atlatl throwing sticks. Some large examples can be interpreted as knives instead of projectiles, or alternatively as tips for thrusting weapons. Small points could conceivably be used on spears thrown by atlatls, but other uses may yet be discovered.

Most Clovis cores are oversized bifaces that also yield large, thick flakes for camp use. Those same bifaces would have been useful as heavy cleavers for dismembering large game and were often split in two for conversion into smaller bifaces and projectile points. The large, platter-like bifaces are most commonly known from caches.

Blades and conical cores are routinely included in the inventory of Clovis technological traits (Collins 1999). We're not talking about delicate microblades, but robust flakes from well-developed cores. Generally there is agreement that, to qualify as a blade, a flake needs to follow a previous flake scar, as part of a deliberate sequence. Another clue to conscious strategy is finding sections of core tops that have been truncated in order to rejuvenate the striking platform and extend the number of serviceable blades that may be obtained. Typically, the blades are 5 to 20 mm thick and from 40 to 200 mm long, with an obvious end-to-end curvature. Despite archaeologists' interpretation that blades help define Clovis use of stone, little evidence of blade use is apparent aside from infrequent caches or at primary quarries. In fact, just the widespread conviction that large blades are uniquely Clovis often leads to their acceptance as Clovis without corroborating evidence.

Blades are relatively common at any period of time, when large rocks are reduced at the quarry. Once a series of blades is started, it is relatively simple for the knapper to follow successively formed ridges and develop a formal core. Without a larger pattern of blade use however, the significance of blade occurrence is questionable. Few Clovis era blades are reported at distance from primary stone sources, and they do not accompany caches of projectile points.

In the far west, occasional crescent shaped tools of unknown use can be connected to the Clovis period in at least one Clovis cache. The fact that the same tool shape is more frequently associated with the Western Stemmed Point Tradition is problematic.

Camp tools are seldom considered distinctive enough to be useful as culturally diagnostic features of a site. However, the consistent use of edge-to-edge flakes by Clovis hunters allows archeologists to arrive at reasonable associations based on how the

Miniatures

Virtually all fluted point styles have been found represented in miniature form. Given the diminutive scale, from a quarter- to a half-inch long, many more must have been made than the percentage of recovery would indicate. Many small projectile points from the Mockingbird Gap site in New Mexico had clearly been used and repaired repeatedly, even well below the conventional estimate of useable Clovis point size. Some rethinking of serviceability is in order. If we can agree that a colonizing group is inherently adaptive, it is not a stretch to imagine points being made at various scales, depending on the size of game to hunt. Some points were made on flakes so small that it is difficult to view them as anything but toys. When we think about how children of any era play, we realize how commonly they imitate the adult world.

Assuming at least some tiny points are toys, we see that the cultural template of what constitutes a weapon tip is well understood among children. Even when flutes are not executed on small points, they are usually represented by a clean flake surface. Overall proportions are maintained, even when there is no functional requirement. At the Jones-Miller Hell Gap site in eastern Colorado (Stanford, 1978), a miniature projectile point was found in ritual context at the base of a large post mold, along with offerings of raven, a small dog, and a bone whistle. The distance between a toy symbolizing adult stature and a miniature symbolizing communication with the spirit world is not at all large. After Paleo-times, miniatures are less common, perhaps because hunting is less central to the lifestyle.

Until we know the significance of miniatures, it is prudent to keep our interpretive options open. While they may indicate a phase of social training, miniatures may represent evidence of a larger range of hafting and weaponry options than normally entertained.

flakes were created. Large cores also lead to relatively thick flakes when compared to later cultures.

Meet the Family

The family of fluted point traditions is widespread and varied, spanning about three thousand years. Before we attempt to speculate on how the technology transformed over time, we should look at a broad outline of some recognized variations of fluted point styles. Archaeologists Julliete and Toby Morrow published a set of morphological trends (1999) that appear to represent a gradual shift in what was considered to be an ideal shape as Clovis people filled the continents. At northern latitudes, biface edges extend only slightly wider than the base and then arc smoothly to the tip. Just north of Panama, bases tend to narrow in relation to the body width and acquire the typical South American "fishtail" basal shape. South American fluted forms are most often stemmed and quite often very thin and flat. From a technological perspective, thinning gained importance in South America, where fewer projectiles were fluted since the requisite thinness was already present. At the same time, there appears to have been a switch from bifaces to large flakes as the starting stage of point production. Early Clovis technology relied heavily on the use of widely spaced percussion flakes that often carried across the entire face of the point. Only minor pressure flaking was used to prepare the surface for fluting.

In the American Midwest, Clovis was followed by the Gainey style of fluting. Gainey points tend to be long, wide and thick when compared to the other late fluting styles, and flutes extend from a third to two-thirds of the point length. Examples of this style are seen from the Lamb cache in Pennsylvania and at Bull Brook in Massachusetts. Bases are parallel-sided and sometimes a second flute is seen on a face. Tip beveling has been reported on preforms of this style.

Barnes points are well-fluted and carefully retouched, but are not so delicate as Folsom. Relatively narrow bases have slightly flaring fishtail shapes, and the basal concavity is usually refined by a special, short flute that leaves a sharp basal edge despite

Fluted point variations

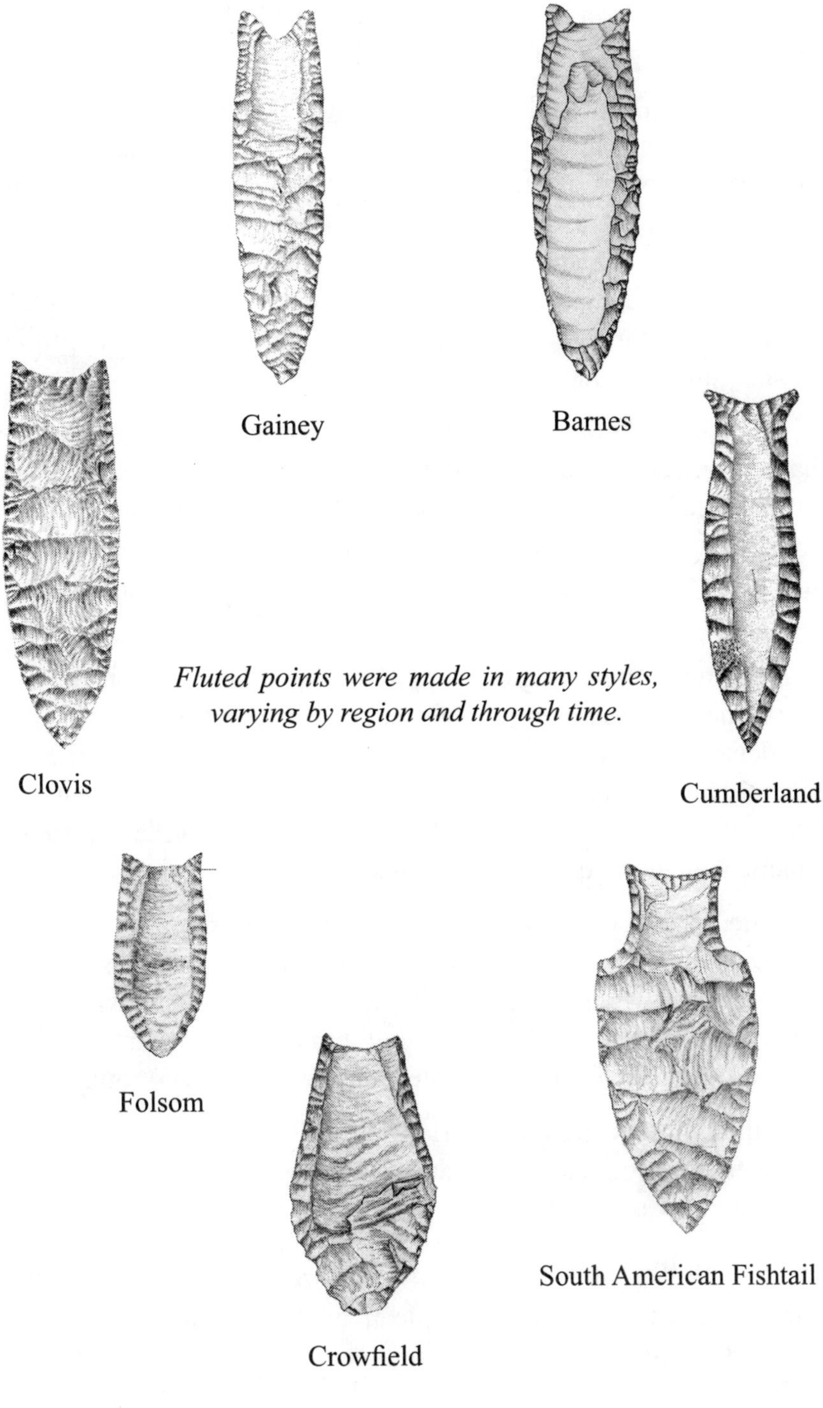

Fluted points were made in many styles, varying by region and through time.

an otherwise thick base. Preform tips were squared or rounded in preparation for fluting, but apparently were not beveled. The Parkhill complex in the vicinity of the Great Lakes is the primary source of information about this point style.

Following Barnes, although rarely fluted completely to the tip, Crowfield points are very thin and wide, with pentagonal shape. A Northeastern point variety, examples of Crowfield points are sparse because they were often shattered in mortuary fires. Holcombe style points are very similar, being widest above the midpoint, and are sometimes compared to "pumpkin seeds." Fluting of Holcombe points declines greatly in importance and is often seen as multiple scars on only one face.

Cumberland points are generally longer, narrower, and thicker than other well-fluted points and have distinctly flaring ears, often emphasized by a constricted base. Most Cumberland points are found in a narrow North-South band along the western flanks of the Appalachian Mountains between northern Alabama and southern Ohio.

Folsom points, dating between 12,900 and 12,200 years ago, are the thinnest of the fluted points, characterized by full-length fluting and delicate retouch. Their width varies very little, and thickness is nearly always under 5 mm. Folsom points occur throughout the open prairie known as the North American High Plains, from Alberta Canada to the border of Mexico.

Fluted points found in South America tend to represent relatively shallow thinning flakes on already thin preforms. The base is typically narrowed after fluting to form a distinctive fishtail-shaped stem. Effective lateral thinning also meant that basal fluting was needed only intermittently, as a last option.

Although there are other variations of fluted points, mostly designated by collectors, they are not yet well enough documented to discuss meaningfully.

The dating puzzle

The best way to settle when people arrived in the Americas would be direct dating. However, locating uncontaminated materials suitable for dating is difficult, and the techniques for dating still need to be improved. A number of radiocarbon dates set Clovis occupation south of the ice sheets between 12,800 and 13,500 years ago, when dates have been calibrated, but claims for earlier dates have yet to be resolved. Carbon 14 dating on individual tree growth rings at the Mesa site (Kunz 2003:19-23) in Alaska vividly illustrates that the assumption that carbon accumulates at a steady rate does not always hold. Supporting data from independent sources like ice cores, coral reefs, tree rings and the like should eventually allow a more comfortable calibration. Until then though, C-14 dates will be viewed skeptically by some. For example, a radical proposal (Firestone 2001:9-16) suggests that a supernova or Tunguska-like low-altitude meteor explosion could have reset the radiocarbon clock over much of North America by adding C-14, contributing to extinction of large species at the same time. An unaccounted infusion of C-14 could mean that calibrated dates are far too recent.

Indirect dating methods using rates of biological change have been proposed that argue for earlier entry than C-14 evidence indicates. Mitochondrial DNA does not accommodate fine resolution but, depending on which assumptions are correct, the dating of first migrations could go back to between 17,000 and 34,000 years. A couple of mutations in haplotype 10, a form of the Y chromosome, lead some scientists to bracket the entry into the Americas from Asia between 18,000 and 15,000 years ago. Linguistic development also assumes rates of change and has been used to argue for up to 30,000 years of development. These methods provide tantalizing possibilities but have not yet been accepted as conclusive.

A number of occupation sites have produced C-14 dates in considerable excess of 11,500 RCYBP. Lewisville, TX came in at 26,000 RCYBP; Meadowcroft, PA dated 28,000 RCYBP; and several sites in South America have reported dates as old as 40,000 years ago, near the limit of C-14 dating capability. The older and more unusual a site, the more rigorous the demand of convincing the rest of the archaeological community that the claims are credible. To satisfy scientific scrutiny, sites must be dated securely, without fear that the dated material has been contaminated. Site preservation also needs to rule out the potential for intermixing by later materials. Isolated finds are not very persuasive. It is much better to see a unique pattern of related technological traits that link the best candidates and demonstrate the existence of a tradition.

Competing routes

The route used by early entrants into the New World has a lot to do with how they might have altered their technology along the way. Land travel is necessarily slow and incremental as people accommodate to new environments encountered as they relocated. We should not expect a quick march into unknown territory. Glaciers, mountains, tundra, and forest each present unique challenges that are likely to have introduced long-lasting adjustments to the tool kit. Ocean travel, on the other hand, opens the possibility, on either coast, of rapid leapfrogging past the most inhospitable areas by following prevailing currents without the need for revising the pre-existing tool kit. With a sturdy raft or boat—large enough to carry a number of people and their gear—one can paddle out to the current and be transported rapidly south. Returning north is much more difficult because of prevailing southerly currents.

Whatever the mode of travel, a threshold number of people is required to spread a population, although multiple incursions could have eventually established the requisite population level. In fact, it is likely that there were many populating waves over the great span of time we are considering. I will limit my concern to which groups are important to spreading specific aspects of technology more than who contributed which gene to the pool.

While some skeptics argue that people did not possess the means to bypass the Bering Strait so early, it is impossible to ignore the fact that Australia was colonized by people who managed to travel 500 miles by sea 40,000 years ago. The Izu islands of Japan were reached by 30,000 years ago, which implies a level ocean-going savvy that might have allowed people to shore-hop to the Americas. Three hundred feet of rising sea levels,

Migration routes

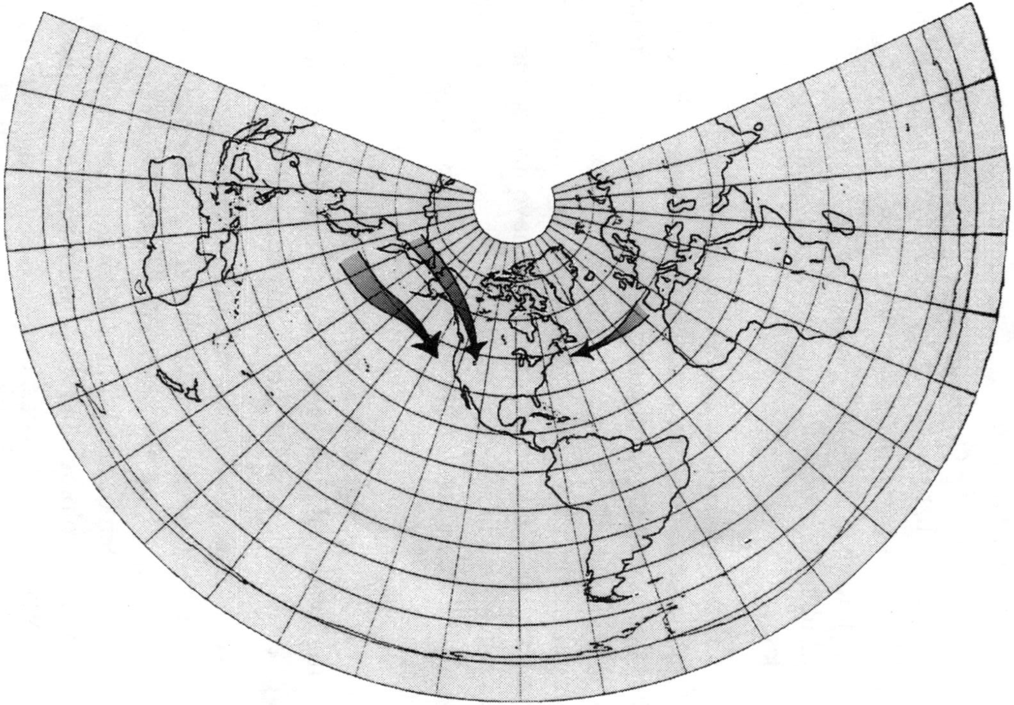

A polar view of routes proposed for early population of the New World illustrates the problems with distance, water and ice.

as well as a lack of agreement on what evidence indicates ocean travel, leave us short of clues.

Placing first colonists on the eastern shores of the Americas presents the same issues as a western entry. It remains for archaeologists to decide whether early American lithic technology could be derived more readily from European roots than Asian ones. For now, the lack of blades as a substantial and formal component of early lithic technology in the Americas is the most persuasive argument that stoneworking did not come directly from Europe. On the other hand, there is nothing about Asian lithic technology that leads convincingly to fluting, and it is as difficult to explain the abandonment of microblades as it is to explain the lack of blades.

It may be that the most tantalizing clues are those most difficult to access. At the same time the Solutrian culture was active on the shores of Spain, an ice sheet covered the Atlantic Ocean from Spain to Newfoundland. Because so much water was locked in the ice, the shoreline on either side of the Atlantic was much wider than it is at present. While people would not have walked the direct route provided by the ice shelf, it would have made a shoreline that would have sheltered watercraft from the usual turbulence of an open ocean.

Similarly to the Beringian entry route, three hundred feet of sea level flooding by glacial melting leaves the most likely shoreline evidence of an Atlantic passage thoroughly hidden.

The ice would have put a most useful food source at the disposal of anyone who wanted to venture there. Like penguins of the southern hemisphere, the great auk was superbly adapted to deep diving in the open ocean. Without natural predators, the auk congregated in large colonies on remote islands, like the Grand Banks off the Newfoundland coast or near Iceland, until sailors of the late 1800's drove them to extinction simply by clubbing them for food, feathers, and fuel. Such a passive and populous prey would have accommodated food needs for trans-continental migration of people.

Contemporary styles

Projectile-point styles are typically listed in chronological fashion, like actors waiting to go on stage. Why not consider that there may have been many players at the same time? Any technology branches and develops in response to local conditions, and only rarely is spontaneously transformed to something entirely new. Completely different, but coexisting, technologies must be viewed as competing for utility in coping with the conditions at hand. As time wears on, a specific technology comes to dominate *a place* for *a span of time*. Can it be that fluting actually was simply a dominant competitor relative to other strategies? If fluting were so superior, why would it then be so completely abandoned later on? To answer these questions, we should recognize that the march of lithic traditions does not always go in single file.

In the far northwest, the separation in time between fluting traditions and the Western Stemmed Point Tradition is slight. Radiocarbon dates on a cache of stemmed points found at Coopers Ferry (Davis and Sisson 1978), in the lower Salmon River canyon of Idaho, comfortably overlap the Clovis period, but do not precede it. On the other hand, a crescent tool typical of the Western Stemmed Point Tradition was included in the Fenn Clovis cache (Frison and Bradley 1999). The only reasonable explanation for Clovis people borrowing a tool design from someone else is that another population was there at the same time. Early stemmed projectile points have no obvious technological traits in common with Clovis. In fact, the stemmed points appear to represent a more settled lifestyle and could be related to stemmed points made at Ushki Lake on the Kamchatka Peninsula of northeastern Russia. The Western Stemmed Point Tradition subsequently provides a relatively steady evolution of projectile forms through time. Even if makers of stemmed points did not arrive first, stemmed points dominated the western basin and range region, which signals a very competent tradition.

On the central Great Plains, a style known as Goshen appears to fill the time period between Clovis and Folsom, but the technology is distinct enough to suggest derivation from another origin than

Clovis although it is not incompatible with Folsom. If another technology was in place, Folsom people could have adapted convenient elements in their subsequent evolution from Clovis fluting.

Early technology along the eastern seaboard includes small blades and thin, flat bifaces. Lithic resources are generally of poor quality or limited extent east of the Appalachian Mountains and may have caused an early shift to accommodate the predominately inferior sources of stone. Alternatively, if a pre-Clovis technology is represented, it may have been in a viable position to spread as Clovis, once better lithic sources were reached.

Ice age environments

It is tempting to look at the comfortable environment we enjoy and assume that it has always been this way. Our brief occupation on this land has been unusually calm when considered in the larger scheme of things and it is useful to look back in time to see what nature was throwing at early hunters. An individual might not have seen anything remarkable about the climate, but successive generations would have encountered enough change that it was not practical to settle for long, and impossible to stabilize their hunting conditions. Those who would like to pursue the potential impacts of environment in detail may want to read *After the Ice Age* by E.C. Pielou (1991) or *Children of the Ice* by John and Mary Gribben (1990).

The Serbian mathematician, Milankovitch, used astronomical observations to show that the earth responds to 100,000-year cycles due to its eccentric orbit about the sun, 41,000-year cycles because of changes in the tilt of the polar axis, and 22,000-year cycles brought about by precession of the equinoxes. Put together, these effects cause rhythmic rises and falls in the solar warmth from the sun that reaches earth. If summers grow progressively cooler, more of the inevitable winter accumulation of snow and ice will be left to build into what can become glaciers.

Typically, the coldest point of a cycle (called a stade) is broken by a burst of warmth. Temperature increases generate water vapor by evaporation that returns to earth as rain. Even as some areas are drenched by rain, shifting circulation patterns leave other areas dry. As the jet stream circles the Arctic in a wavy pattern from west to east, many variables, including mountains or ice sheets, can favor one side of the globe with mild weather while causing the opposite on the other.

Carbon dioxide levels amplify effects of the Milankovitch cycles. When oceans are more productive, more carbon dioxide is drawn from the air and deposited in ocean-floor sediments. Earth during its ice-age was both colder and dryer than at present, allowing iron-bearing dust to be blown into oceans where it was consumed by plankton. The biological use of iron by plankton accelerates cooling by reducing CO_2 in the atmosphere, but also encourages more warming when the dust supply is cut off. In a similar chemical cycle, weathering of silicates takes CO_2 from the atmosphere, while ice sheets retard the depletion.

The melting of ice sheets 14,600 years ago marked the onset of what is called the Bølling/Allerød climatic phase. When the ice sheet melted, as evidenced by ice cores from Greenland, half of the warming took place in only fifteen years. Archaeological evidence from Spain shows a catastrophic environmental impact in Europe at about the same time. Prior to melting, the Canadian ice sheet stood about a mile high from ocean to ocean and blocked the jet steam from pulling frigid arctic weather southward. The resulting climate, though 6 to 8 degrees Fahrenheit cooler than now, caused less severe winters. Without strong seasonal fluctuations, ecological zones of the north intermingled with those of the south, creating conditions for lush, tall grasses on the North American plains and mixed conifers and hardwoods in an open mosaic through the Midwest. No comparable ecology exists anywhere on earth today.

The general drought that accompanied the Allerød phase was eased from 13,360 to 13,100 years ago by the Intra Allerød cold period, although the terminal Allerød brought even greater drying with the "Clovis Drought" at 13,000 years ago. Following that, a catastrophic temperature drop, cold enough to freeze remaining surface water introduced the Younger Dryas phase. No mammoth or Clovis remains date later than this freeze. For the next 800 years, increased moisture and continued cold dominated the climate experienced by people using fluting traditions.

Simply relating the chronology of a succession of glacial advances and retreats fails to impress the true scope and magnitude of ice ages on the earth's inhabitants. Some scattered examples may help us better appreciate the impact.

Between 15,000 and 10,000 years ago, sea levels around the world rose by 50 meters, then by another 40 meters by 5,000 years ago to reach near present-day levels. Not all the released water went directly into the ocean. Much of present-day Nevada was covered by Lake Lahontan, which encompassed 21,000 square kilometers of the Great Basin, before its level fell near 14,600 years ago due to evaporation. Lake Bonneville, covered 50,000 square kilometers including the present Great Salt Lake to a depth of 300 meters before it found an outlet at 14,500 years ago. Down cutting opened a channel sufficient to provide a flood volume six times as large as the maximum flow of the Mississippi River, leaving Utah's present Great Salt Lake as its last remnant.

Impressive as these amounts sound, they pale beside several lakes that formed by natural dams along the southern ice fronts. Lake Columbia and Lake Missoula were created before 17,000 years ago by ice dams that periodically floated free to cause great floods across Montana, Idaho, and Oregon. These huge glacial outbursts, known as *jokulhaups*, occurred over 40 times until the cycle broke, 14,800 years ago. The collective outflow of nearly 300 cubic miles, roughly six times the volume of modern Lake Ontario, scoured the land each time in less than two weeks, with flood depths of 800 feet.

Starting about 14,000 years ago and lasting 4,500 years, Lake Agassiz and Lake McConnel covered 350,000 square kilometers of the southern Canadian landscape with a volume of water over twice that of the current Great Lakes. The first outflow fed the Mississippi River but subsequent drainage flowed through the St. Laurence Valley and lastly into the Hudson bay. Those shifts spread fish species over vast areas of currently unconnected drainage. The final outburst happened as recently as 8,200 years ago and sent so much fresh water into the Atlantic that temperatures in Europe and Greenland dropped for about 200 years (Clarke, et al, 2003:922). Successive generations of people, game animals, and biota had to adapt to the cumulative alteration of much of North American landscape.

Megafauna

The discussion of lithic hardware must take into account what people were able to prey on. Before humans entered the New World, the land was occupied by an impressive array of now-vanished megafauna including mammoth, mastodon, giant sloth, camels, saber-tooth tigers, cheetahs, American lions, short-faced bear, giant beaver, and more. The extinction of such a panoply of creatures, including 34 species, in a very short time leads some to the conclusion that early people, in their march of domination over two continents, were overly efficient agents of destruction. Despite an appearance of guilt by association, no conclusive evidence has been found to link the demise of species to hunting by man. Many of the now vanished large species have never been found with artifacts from human predation. Other species that disappeared were too small to have been significantly pressured by human hunters. Another suggested reason for extinction is disease that may have been introduced when humans entered the Americas, but no direct evidence for it has yet been found. There were probably many causative factors, like most other catastrophes.

Extremely rapid climatic changes occurring at the same time, when melting glaciers allowed the jet streams to strengthen seasonal contrasts, almost certainly squeezed some of the larger animals into untenably small ecological niches. This, in turn, pushed people to deal with new environments and greatly different game habits. Prior to the climate change, the American landscape resembled lush African savannas in the quantity and variety of game. Large fierce carnivores may have kept human populations in check before the end of the Pleistocene. There is no reason to believe that people of that time were equipped to take on such predators.

Carnivores cause damage to bones in a way distinctive from humans. Even the biggest predators went for the soft parts at the ends of long bones because that gave them nutrition without endangering their teeth. Predated bones have tooth scrapes and punctures—not impact breaks. Humans, on the other hand, are after the well-protected and highly nutritious marrow, and break long bones at the middle to get at it. Furthermore, fresh, "green" bones break with characteristic spiral fractures when impacted, while dried bone breaks relatively straight along the length or breath of the bone. Hard hammer impacts on bone also develop a series of nested conical fracture surfaces, very much like cones seen when hard projectiles impact plate glass. These distinctions are quite important for archaeologists pondering potential human activity at paleontological sites.

Since a number of sites reveal that Clovis people made use of a wide variety of animal resources, it raises the question "Why are Clovis people so strongly associated with mammoth?" Most sites that show a clear linkage were found because the large bones drew attention to the site, and artifacts were discovered later. Many of those sites were noticed by paleontologists looking for fossils rather than by archaeologists. Important clues about the nature of Clovis habits may have been overlooked because of poorly founded assumptions about where to find evidence.

Settlement patterns

In the eastern US, most quarry locations have significant Clovis camps nearby. There is some speculation that Clovis sites are more abundant in the east because that is where the technology started. Clovis sites in the west usually represent short stays at a kill, and indicate a pattern of unrestrained wandering rather than long term occupation of restricted territory. The Mockingbird Gap site in New Mexico is the only Clovis site in the Far West where groups might have congregated and exchanged resources (Weber 2001). Debris at Mockingbird Gap bears a striking resemblance to that seen at large Folsom camps in the same region, and may represent the transition between Clovis and Folsom.

Following Clovis time, sites reveal larger concentrations of people. The Lindenmeier Site (Wilmsen and Roberts 1983) is a huge Folsom base camp in northern Colorado that apparently was utilized repeatedly. In the northeast, Vail, Maine; Debert, Nova Scotia; Shoop, Pennsylvania; Gainey, Michigan; and Nobles Pond, Ohio all show signs of repeated visits or extended occupation. One possible interpretation is that these sites represent people in the process of adapting to local conditions and on the verge of transforming their technology. Scheduled rendezvous sites may have accommodated necessary social exchange in the same manner that early trappers met yearly for supplies.

Acclimation

Settlers of any land take hold very quickly. Even if we accept that Clovis people were the initial inhabitants of the Americas, it is highly unlikely we could find remnants of the very first colonists to arrive. Any archaeological evidence, no matter how early, must have been left by people already adapted to relatively local conditions. It is naive to suggest that early hunters lacked familiarity with resources. Early pioneers of the American West took less than a generation to learn how to get along in a new setting. In each case, some aspect of the landscape serves to define a scope of land use. On the high plains, the extent of unbounded territory is staggeringly large and where game went, people followed. To the west, mountain ranges and river valleys defined relatively restricted parameters. In the east and Midwest, original environments have been altered so much it is difficult to estimate what may have defined territorial limits. Makers of Dalton points appear to have used river systems as the framework for a durable, long lasting lifestyle immediately following Clovis times.

By looking at how durable resources such as stone are distributed, territorial delineation becomes more evident over time. Some territorial domains remained in place even into historic times. Others shifted as ecosystems changed or as exploitation of new resources created new boundaries. The fact that Clovis people moved stone up to several hundred kilometers illustrates the scope of their geographic knowledge base.

Depending on the degree of utility that the original technology afforded with each newly encountered territory, the shift away from Clovis patterns was sometimes rapid and sometimes slow. People that lived in relatively closed-off environments in mountains and along coastlines were quickest to adapt to local conditions. Slowest to change were nomadic hunters who utilized the broad, stable ecosystems of the plains. Even there, as animals that depended on long grasses were replaced by those that thrived on short grass, people had no choice but to respond. Where forests encroached on once open savannas, the technology changed as well. The earliest territorial frameworks were probably formed to exploit specific animal resources. Mammoth range is known only from rather indirect means, but is linked to Clovis occupants. Later people used fluted point technologies to support their lifestyle by hunting bison or caribou. The eventual shift to less mobile game was accompanied by a more opportunistic pattern of stone use.

Conclusions

If it seems as if there is no particular conclusion to be reached from the evidence at hand, it is because archaeology records scattered sites representing independent activities of early people. Without a connective framework, opinions often conflict. About all that seems certain is that people arrived in the Americas by the end of the Pleistocene, and encountered extraordinary and rapid shifts of climate. Fluting was established before mammoth disappeared and was abandoned after about a thousand years of use. Phrased this way, the prospects for making sense of the past seem dim. As the issues highlighted in this chapter are further explored however, concepts will be developed that allow an order to emerge from independent data. Despite apparent confusion, archaeologists have provided a wealth of information essential to shedding light on the past. By considering the evidence in an appropriate perspective, the history of fluting can be deciphered.

Chapter 3

Transforming Technology

If we are to make sense out of the progression of lithic technology over time, we must consider what change involves. Today we may see change as unavoidable, but that has not always been the case. As contradictory as it may sound, changes in the way we do things are usually intended as a means of creating stability. Sometimes stability comes from maintaining the status quo but, in a crisis, it may require departing from the familiar. We will concentrate our attention on a particular stretch along the road of technological transformation, when fluting was practiced. The course people took was of their own choosing, although events steered their decisions toward an increasingly narrow range of options. Before proceeding directly to the discussion of technology, it is important to appreciate the backdrop of social trends that influenced the acceptance or rejection of new ideas.

Role of tradition

When specific elements of technology are used by successive generations in a particular geographic area, that constitutes a tradition. Tradition represents the way a group of people decides to use certain techniques to standardize their routine tasks. Highly traditional people live by rules that encompass every aspect of their lives. Decision making is easier when there is a rule to cover every contingency. Inevitably, though, changes beyond the tradition's control, like a new climate, make the old ways falter and new rules must be created. The way traditions evolve to incorporate new requirements is the subject of this chapter.

We cannot know for certain how strict the societal rules of over 10,000 years ago were, but consistent patterns in the way artifacts were made testify to the importance of tradition. Variations in those patterns over time prove that traditions were forced to adapt to successive changes. Some traditions were stable for thousands of years, while others lasted only a few generations. The musical, *Fiddler On The Roof* is about people who are being uprooted from their traditional way of life. That's not a bad analogy for both the inevitability of change in early times and the difficulty in facing the consequences. Instead of pogroms, prehistoric people faced unforeseen problems as diverse as climatic change, scarcity of resources, and influx of competing populations. To understand why certain patterns of behavior developed, it is helpful to consider what problems were thrust upon the group and what resources were available to deal with them. Although this book is specifically concerned with lithic technology, other components of culture, from textiles to food gathering to religious observances, were equally subject to change.

It is inevitable that we should form opinions about past cultures based upon our modern frame of reference; such comparisons help us empathize with peoples of the past. But we must keep in mind that their lives and experiences were very different from ours. One route to reconstructing the past is to extrapolate from the ways present-day people cope with similar challenges. Since modern people have a unique history of sociological development and differing basis of technology, it is difficult to demonstrate that their decisions are the same as other groups might have arrived at in the distant past. Simply declaring what human behaviors are shared, regardless of culture or time period, is fraught with even more danger of dispute.

In today's world, deep traditions often look old fashioned to our modern eyes. As values and priorities shift, it is increasingly difficult for traditions to remain current. When a tradition fails to keep up, it becomes an island of past attitudes. Modern "primitives" such as African Bushmen or natives of New Guinea may look to us like good candidates for comparison to paleo hunters, but in actuality, they have developed their traditions under their own special circumstances over thousands of years. A way to circumvent the problems of using modern analogs is to look for evidence of persistent behavioral patterns in the archaeological record. By correlating traditional tool usage and design with living conditions of the times, we can begin to differentiate cultures of the past.

Although traditions are the foundation of archaeological description, they can only represent periods of relative stability between episodes of profound disruption. The short span of re-invention is most apt to be invisible archaeologically. In that case, comparing the traits from the end of the earlier tradition to those at the start of the later one can provide clues to what spurred the change. Much of why a tradition undergoes changes through time is connected with fluctuations in nature. Wet times require different coping skills than dry ones, grasslands offer different resources than forests, and animal herds can change their locations overnight. While the surrounding environment sometimes changed, at other times people's travels forced them to cope with a new environment.

Individuals working together to accomplish a common goal form a social system whose knowledge base is most visible in their technology. Ideally, the technology is well aligned with the environment, but since environments always fluctuate, any group eventually finds cause to adopt new ways of doing things. Change occurs most readily when people share a common perception of their needs. An outsider may have a workable solution, but if it is too "foreign," it will likely be discarded. Popular change means that people share a positive view of the new ideas. Social learning theory holds that individuals in diffusion networks frequently learn from each other by observational modeling. Demonstrations make technology more visible and understandable for another to copy. Typically, demonstrations are very effective at speeding the rate of diffusion. The deeply seated practices of lithic technology are admirably suited for such transmission of technical knowledge. Training was not so much a task as an integrated part of life, and children could have learned simply by watching how adults accomplished their jobs.

Technology can be viewed as a design for action that reduces the uncertainty of an outcome. Typically, technology has aspects of both hardware and software. In other words, to use the tools correctly you need clear operating instructions. Everett Rogers (1995) lists five criteria that allow an innovation to succeed: A *clear advantage* or value must be evident in the proposed change. *Compatibility*, or how change fits with prior practice, minimizes disruption. Complexity interferes with the ease of comprehension, so *simple ideas are easiest to sell. Solutions that can be tried without first discarding a previous practice* also facilitate adoption without risk. Finally, people need to see the solution *succeed in practice.*

The fact that fluted points exist means these criteria were successfully met. Later modifications also had to pass through the same filter to be accepted. Decisions on whether or not to change progress through stages: Unless a strong need already exists, people can be slow to recognize options beyond those they are already familiar with. Successful persuasion to change usually requires evaluation of the presumed advantages. Trials and demonstrations help expedite the decision of whether or not to continue with the change, and implementation is considered complete when the idea is no longer seen as new. At that point, ownership has been established and the solution is embedded in normal behavior.

The fact that people used different technologies at different periods of time gives archaeologists a way to differentiate remains of one group from those of another. We need not worry for now about how dates are assigned. Our emphasis will be on how what was done at one time affected what was done at a later time. By noting which elements changed, we can hope to understand how the inhabitants of an area were pressed to make those changes. Knowing how tools were used to cope with changing needs is crucial to understanding why the tools themselves changed.

Tools go out of service for a number of reasons. Sometimes they were discarded in nearly the same condition as when originally fashioned but more likely they were re-used, re-sharpened, and re-built into new tools. As successive generations made and maintained tools, subtle alterations in design came about to accommodate new needs. The changing morphology of many tools serves to distinguish artifacts of one time period from those of another.

Unfortunately, nature mounts a constant attack on physical material left by humans. Wind and water move and erode anything in their path. Since organisms and acids in the soil dissolve and eat away all but the most durable and protected remains, organic material of any kind is rarely recovered in good condition. For that reason, stone becomes one of the most important sources of information about the earliest cultures. Stone implements accurately record how lithic technology was used and, when we understand the sequence of change, mark the time of their use.

It is virtually impossible to recover more than a fraction of the materials that were originally used. The prospect of truly assessing the level of technology from a partial inventory of goods is daunting, but still necessary. Fortunately, human behavior causes people to select what is really important by storage, curation, and caching. As more sites can be found and examined, it is increasingly clear which items held special value for a tradition. Each area of technology can reveal a different picture when considering levels of community. Thus, it is important to look separately at hunting, communication, transport, art, commerce, religion, agriculture, travel, scope, clothing, storage, weapons, shelter, music, excavation, temperature, illumination, ... well you get the picture.

In our quest to understand the lineage of technology, we must keep in mind that technology is mankind's means of escaping the constraints of environment. Every addition to our technological baggage alters our environment in a cumulative way, leading to far reaching environmental consequences that are nearly always unanticipated. Response to those impacts results in an ever expanding web of cause and effect. Eventually, the artificial environment we build around ourselves is of itself the predominant stimulus for change. Cell phones are compatible with our material world, but useless to primitive people without electricity. So, understanding the context of a technology is essential to our interpretation.

Earliest cultures had relatively little technology to shield themselves from the rigors of a wild environment. Limited material inventory can be expected to allow mobile lifestyles to remain reasonably stable because adding to the material load is not practical. Technological innovation is generally promoted by settling because adding equipment is less burdensome when it does not have to be carried. So, the more settled you are, the more you can accumulate. Specialization affords a way to maintain diverse equipment, as long as survival does not depend on knowing how to use every tool. We can appreciate what diversity brings by looking at how we live today compared to any other period of time. Advances in technology inevitably accompany cultural expansion.

To appreciate the level of technological accumulation, it helps to compare different levels of community. The activity that people engage in depends strongly on the level of community they are supporting. When the material associated with an individual is not noticeably different from that of a family or clan or cultural tradition, we can conclude that social or political issues had relatively little influence over technology. Individual expression will likely be suppressed until consensus can be reached, because the action affects everyone. What you do as an individual may appear to be independent of everyone else, but if you are feeding a family, you are no longer able to act strictly as an individual. That hierarchy extends upward to clan, tribal, state and ultimately global responsibility. Divergence between low and higher levels of community tends to encourage innovation by individuals that is sometimes even commissioned by leaders. Recognition of the differences in technology as it operates in support of the various levels of community affords great insight into ancient behavior.

At simple levels of technology, the impetus to do things the same way they always have been done is strong because those ways have stood the test of time. Change tends to occur only when something outside the control of a group forces an accommodation. Individual innovation in early cultures would have been vigorously resisted because of the strict requirement for broad consensus. Significant environmental upheavals like those at the end of the Pleistocene may happen only a few times over hundreds of thousands of years. But when they happen, they can be so severe that they force radical technological responses. Social change is more often subtle and can transform technology gradually from generation to generation as improvement is accepted by small increments. By widening their scope of sustenance and becoming less bound by tradition, later cultures made standardization increasingly less important and fostered individual innovation that was more readily incorporated into the life of the community.

Presumably, big-game hunters had a relatively restricted definition of community because they were necessarily dispersed. While they did not carry the seeds of change through social and technical advancement, they were still subject to the whims of nature. Because of their limited community structure, their technology was most tuned to coping with their physical environment.

Adjusting to new conditions of game or climate can be as simple as changing the size of a projectile tip, or as complex as designing an entirely new projectile form. If the plants used for spear shafts no longer grow because of climate changes, replacement shafts are likely to demand new kinds of tips. Greater stresses, such as climatically induced migrations to a new region with unfamiliar resources, are likely to cause fairly radical adaptations to the tool kit. Gradually-expanding populations force accommodation to restricted range and resources that are often reflected in technology. Whatever the reason for change, it is reflected in the archaeological record.

Folsom technology appears to follow a broad outline of tool production used in Clovis times, even though the environment and game of choice changed drastically. The next change in projectile design, to Agate Basin points, bears no obvious relation to Folsom, although the environment apparently was much the same, only drier. This shows that that the effect of environment may not be conveniently straightforward. On one hand, an entirely new end product may not require great changes in technology. On the other, it is possible that Agate Basin technology was brought in by newcomers. Therefore, any change in technological expression should be questioned for its environmental or sociological trigger. Because tradition reflects attempts to manage the application of technology, its manifestation gives archaeologists a powerful analytical tool.

Process-step controls

Every part of a process has its own controls that ensure a reliable outcome of the task. The qualities we think of as representing the essence of a stone tool type come from controls incorporated into the production process. By attending to the success of each step of the process, the emerging tool can near the limits of what is physically achievable. Process-step controls can be seen as the foundation for tool-making recipes. Relationships between tool types may be indicated by how many process-step controls they have in common.

Industry recognizes that product quality rests on removing variations in the production system. Quality is achieved by instituting controls over each step of the process, which allows the tool to satisfy the functional requirements of a user. Controls can involve equipment that standardizes output, or specifications that reflect requirements.

Many process controls deliberately use specific conditions that lead to highly predictable results. However, telling a flintknapper that an angle of 27-degrees is optimal does no good if he can't swing his hammer that accurately. Of course you can adjust your sitting position, or build a mechanical device, or devise a template that deflects all errant blows. Any of these actions makes sure you do the right thing each time to get the same result each time. Process-step controls necessarily limit the conditions available for fracture and raise the prospect that they can be detected as specific signatures in the debitage. The fact that different groups may choose their own unique set of process controls makes it possible to recognize their work as distinct from another.

The preference for a particular sub-set of all possible options may be arbitrary, or it can be chosen as a process control factor. For example, Folsom fluting platform locations vary much less than they theoretically need to. The particular specifications Folsom people chose allowed them to use the same technique for each fluting attempt. In this case, controlling the platform location is much easier than controlling precise angles of force.

Without controls, modern flintknappers tend to resort to shortcuts and use skillful improvisation to correct defects that invariably creep in. They also tend to copy a variety of styles and rarely have to use their points to feed themselves. Consequently, only a few modern craftsmen can match the standards set by ancient people, without using grinding wheels or metal tools.

The notion that relying on extraordinary manual skill is largely a detriment may be difficult to accept, but it is useful in explaining long-lived traditions. Individuals of any group at any time vary greatly in their aptitude. If their talents were given free rein, there would be too much variation for typology to be useful. Using the same control mechanisms over and over allows an apparently complex set of process steps to be executed with reliable precision by the average worker. Detailed records and databases were not needed to see what worked and what did not. Practicality and common sense led people to adapt and improve their processes incrementally.

Craftsmen naturally strive not to repeat mistakes that make tools function at less than fully efficient levels. They also are quick to note when an inadvertent change produces favorable results. This model of evolving process steps is slow but reliable.

Variations in finished products reveal the standards of the makers. Even with controls, variations in an artifact set remind us that there is necessarily a prescribed level of tolerance. Discards likewise hint at the limits of tolerance. How closely a feature of an artifact adheres to standards may speak of its importance to the group that made it, or it might reveal a physical constraint. For example, the fact that bases of Folsom points are remarkably consistent in width leads many investigators to suspect that a hafting requirement was responsible.

Stages of work

A tradition arranges techniques and concepts into an order of business which is seen in archaeology as stages of work, from quarry activity, to making tools, to repairing damaged tools. Each tool form can be viewed as having a specific and often unique recipe for construction. While there are many options for how to achieve each goal, settling on a specific set of techniques makes it possible to develop a standard, repeatable series of actions that work together to construct a tool. Each sub-task required to make an implement is a process step. Once an order of business is entrenched, it tends to be difficult to displace because generations of refinement and training build in success.

Grouping sets of process steps facilitates separating work into logical stages of activity. It can be difficult for the worker to adjust to differing muscle requirements from one activity to another. Since motor habits that can be repeated are easiest to apply consistently, a worker will often work on several items at the same stage to minimize waste. Percussion, for example, is most efficient when its rhythm is not interrupted. When work is resumed at the next stage, the craftsman can be more effective by focusing only on that stage.

It is important to recognize that staging is a natural division of the production process, and not an arbitrary devise for sorting artifacts. Even when routine work is conducted from start-to-finish at a location during a single session, stages of progress tend to emerge when tools or work patterns have to be shifted. Whether advancing from percussion to pressure, or using retouch flakes rather than thinning flakes, a new stage of work is defined. Considering why the stages appear can make order of otherwise jumbled evidence. One motivation for staging comes from keeping intermediate products in reserve until they are needed for a particular task. By not committing the tool's use too early, more options are available. So, if the next need is for a knife rather than a projectile, it can be met without dismantling yet another preform. A less obvious impetus for staging is that flakes themselves are sometimes the most valuable products. When flakes are valued as highly as the biface, it makes sense to detach them only when the need arises, otherwise they may be lost or damaged.

Primitive peoples stocked their tool kits with appropriate numbers of each item, according to the needs their tradition anticipated. As tools were used up, preforms were advanced to the next stage of production and new raw stone was introduced at the beginning of the process. This behavior kept appropriate tools readily at hand, even far from a source of stone. Some preforms may have doubled as cores or eject-a-blades to satisfy the need for occasional expedient flakes. In stone-poor areas, it may have been rare to make a tool from start to finish in one sitting, unless the tool kit was exhausted or the tool blank was too small to yield usable waste flakes. Rather than committing a blank irrevocably to a specific task, it is valuable to bank the option of diverting to another tool form. Staging work activities encourages different parts of the process to be done at different times, possibly by different people. Specialization can be difficult to recognize, but it tells a lot about the social organization of a group.

Tool caches are particularly useful for recognizing broad outlines of production sequencing and staging. Caches provide valuable snapshots of stages of production interrupted in a freeze-frame that clarifies at least those particular phases. More specific insight into process steps comes from analyzing work that was lost or abandoned in progress. Eskimo tradition obligates a user to replenish the cache at the next opportunity, which means that the quantity of material stored is related to the level of anticipated need. Experience shows how much of each product will be required for the group to survive comfortably, without being overloaded. Archaeologically recovered caches teach us a great deal about what the makers valued and what kind of staging was used.

Dr. Bruce Bradley (1996) observes that certain critical process steps, which he calls "technological necessity," must sometimes be attained for a job to be continued. For example, both Levallois flakes and Folsom fluting depend on acquiring specific surface morphology before the succeeding phase of operation has reasonable chance of success. A step that must be accomplished before remaining steps can proceed marks a unique stage that helps define a technology.

Strategic concepts

A worker's action can be wasted if he does not have a clear plan to guide it. While we can readily see the result, the strategy behind the act is much more difficult to recognize. Strategy can be incorporated as a process control, but it should also be acknowledged as a culturally embedded way of thinking. When patterns of thought are thoroughly ingrained, it takes a radical leap of intuition to recognize an alternative mode of behavior as possible. Even then, before change can occur, the whole group has to switch their thinking.

Once value is attached to a part of a technology, it is hard to overcome. The preferences do not have to be rational or even reasonable to be important. For example, a stylistic preference for curved shapes can block the use of angular designs.

Some concepts are so basic that they shape nearly every decision by a craftsman. For example, in order to get consistent results while knapping, you need to know the right place to hit the stone. Once a mental template is adopted, other impact-site options are difficult to implement. Even when you can imagine what to do, your muscles may be too well trained in another habitual response to cooperate. As a result, in technologies where tools made from conical cores satisfy most needs, bifaces tend to be poorly developed.

Because products of mutually exclusive strategies are relatively easy to distinguish from each other, such contrasts of strategy help focus our attention on the really meaningful points of departure among technological traditions. As an example, the kind of preform support that is normally used determines whether a flake can reach full length or not. So, when overshot flakes are present, it is safe to say that the workers relied on good support of their preforms. At the same time, relaxed support is not likely to be part of the same tradition. Other exclusive strategies may show up in early biface morphology and impact spacing.

Incremental innovation

Once process steps evolve, they tend to remain stable. Knowledge is not abandoned carelessly and strategies that have been found to be successful are more likely to be repeated than to be altered. The principle of incremental innovation depends on having stable primary processes. Even though a process step is stable, inevitable departures from the norm may be recognized as improvements and eventually incorporated as a new process step. By altering a process in tiny increments, it can become something the originators would have trouble recognizing.

To survive in marginal environments, people rely on being able to make serviceable tools without fail. The fact that we are able to recognize traditions of technology that persist for generations, suggests that there probably were taboos against idle alterations of tool designs. Changes generally are as conservative as possible to minimize both potential error in manufacture and the effort of redesign. As shifts in subsistence strategy lessen the danger that an untested alteration could bring disaster, greater tolerance can be expected for innovations.

New process steps usually expand from existing knowledge. The earlier in the process a change occurs, the greater number of process steps will have to be re-designed. It is not reasonable to throw away everything your ancestors taught you, and it's not smart to abandon what works before you have a good replacement. Once bifacing was found to be a useful way of providing both flakes and tools, it reoccurred through a succession of cultures. Through time, the basic means of breaking rock apart at early stages of work are followed by increasingly sophisticated stages of refinement. Habits formed from experience in the previous process are hard to shed and process steps that have stood the tests of time are difficult to improve on.

Quantum jumps to totally unheard-of approaches can happen, but are rare compared with additive ones. As a consequence, technology tends to mature gradually. People gathered data through repeated experience, accident and trial, not because they understood the physics. By doing what worked, they gradually progressed towards unknown prospects.

Limiting constraints

People can seldom find a perfect solution to any problem. No resource is universally available and no attribute is perfectly independent of another. People are always faced with restrictions and forced to make trade-offs between competing solutions. Sometimes the decisions pay off, but often they must be reconsidered when they don't pan out. No two cultures find the same path through all the variables open to them, so what we discover by archaeology is unique for every past population.

The best ideas will languish if the wherewithal to execute them is lacking. DaVinci's design for a helicopter had to wait hundreds of years for the right components, to turn a brilliant idea into reality. We can easily lose track of the difficulty in finding suitable supplies of raw materials and tools to transform them in order to maintain a technology. Modern knappers may find copper to be a great knapping tool, but it usually wasn't readily available to early people. Similarly, a limited source of superb stone may not be sufficient to support the needs of a widespread population. Along the eastern seaboard of the US, most cultures depended exclusively on dominant low-grade stone while completely ignoring a small amount of local high-grade rock, because the technologies were incompatible. Such limitations may restrict the technology or trigger an effort to overcome the constraint.

There are instances where limited availability of stone can be adjusted by heat-treating the inferior grades of local stone, or trading for materials from other locales. It can be reasonable to use non-local sources if visits to remote quarries can be integrated into routine seasonal activities. The effort that is expended to acquire raw stone, whether directly or not, provides a measure of the value attached to that particular resource.

Theoretically, there is a vast range of ways to break stone. However, people used only a limited repertoire based on their physiology (how they are built) or sociology (how society conditions them to think). Regardless of the basis, those preferences make it possible to distinguish the work of one group of people from another. Evaluating where potential modes of fracture were avoided requires understanding the complete theoretical range of knapping practice. The constraints of using a limited set of knapping possibilities have a lot to do with what people could make or not.

Nature selects shape, weight, density, and strength of knapping implements. Geography and environmental constraints then determine whether stone or antler hammers are available and how effective they are. For early craftsmen, it made sense to take what was ready at hand in preference to putting a great deal of effort into modifying tools. Trade and travel had the potential to bypass local constraints.

Social mechanisms such as religion or taboo may keep certain techniques in use and bar others. Whether the reasoning is based on fear, superstition, or practicality, group decisions that are reinforced by stigma or threat of harm are powerful deterrents. As we look back in time, we need to remind ourselves that our rationality may not reflect those of an ancient people.

Behavioral aspects, like how you swing a baton, have much to do with anatomy. Depending on how you are used to sitting, it can be awkward to use certain hammer strokes. I usually sit in a chair, and have trouble trying to knap while sitting on the ground. Varying a knapping tool or changing an angle of approach may require the craftsmen to manipulate tools in a new way. Each decision takes the craftsman down an alternate path of behavior. The question becomes one of whether behavior is that of an individual or shared by a group.

Although individuals always exhibit varying skill levels, cultural tasks must be designed to be in reach of every member. Process controls are useful in making technology available to the broadest spectrum of the society by using rote to guide craftsmen. Following a set of rules can substitute for amazing ability. Therefore, at the lowest levels of community, manual skill must be subordinate to process-step control. Particularly skillful renderings of an artifact should not be automatically extended to reflect high levels of skill at a group level. In fact, it can be argued that atypical projectile points result from skillful adaptation. At higher levels of community, skill becomes a highly valued privilege of the elite. In other words, the commission goes to the best-qualified artisan, not to just anyone.

When process-step controls are used to good advantage, it is possible to obtain impressive results that appear to come from skillful knapping. Examples include Folsom fluting, Danish daggers, and Egyptian knives. While each of these items regularly were made with machinelike precision, they were also made in too great numbers to explain them solely by skill. The high levels of standardization also argue against individual artistic expression that normally accompanies highly skillful production.

The more important a task is to survival, the more critical it is that the tool be suited to the task. People with minimal tool kits are much more dependant on their utility than people with a large selection of tools. Survival depends on finding what *works*. The usual phrase is "form follows function," but it is probably more appropriate to say "form *facilitates* function." When a form was seen to present a useful advantage, it was exploited. Likewise, an action that produced a favorable result would have been noted and readily incorporated into the technology. This does not mean that every group solved problems the same way, only that their solutions satisfied their needs.

Technology must fit lifestyles

Increasing technological options can interfere with mobility. Equipment is designed with a normal range of use in mind. If the equipment can't be easily repaired or replaced, it may be unsuitable for mobile applications. A nomad won't be willing to lug cumbersome equipment when a simpler tool set can do the job. Sedentary lifestyles can allow supplemental technologies like basketry and pottery to flourish, with the effect that lithic technology becomes less and less important.

Since innovations are adopted because they prove to be indispensable, they tend to reoccur in solutions to other problems. Many modern flintknappers advocate mechanical fluting mechanisms because they have learned that leverage gives them extra control while breaking stone. While it is tempting to extrapolate from our own social climate, there is little reason to believe that leverage could have been an important component of aboriginal fluting technologies. Until people settle into permanent communities, mechanical devices were too costly in terms of transport and maintenance. Furthermore, if leverage were that useful, it should have been used to address enough other problems that we could see it in material recovered archeologically.

By using highly structured processes, fluting traditions were able to create tools that are difficult to duplicate by even the most skillful replicators. Later peoples gave themselves more flexibility by adopting more generalized means of making their tools. Consequently, it becomes more difficult through time to distinguish tools of one tradition from those of another. The paradox is that as groups see themselves as being less like each other through time, their technologies may become more alike because of compromises between alternative solutions.

When technology develops too extreme a pattern, it is in danger of being incapable of responding to new needs without drastic modification. Fluting appears to be an example of highly specific conditions for tool production that were too constraining to persist.

Drawing conclusions

Development of lithic technologies is somewhat analogous to biological evolution. Bones are shaped by the loads they are required to bear in life. Protrusions on the bones show where ligaments attach and indicate which muscles are strong or weak. Overall size, shape, and density of bones reveal much about whether the animal's environment is terrestrial, aquatic, or air. Evolution acts as an engineer to accommodate whatever niche the animal occupies. Knowing what causes bones to be shaped as they are explains how they vary from era to era. Unlike biological evolution, human decisions influence changes in stone tools, but the same general principles apply. Once we know how a flake is made, we can begin to understand how people gradually moved to different ways of making flakes.

Because stone tools were used to cope with every aspect of early life, their development can be considered to reflect at least some of the history of the toolmakers. As we describe morphology, scar patterns, and stone preference, there should be an underlying linkage to the evolution of the tool. Processes exist because they were passed from person-to-person and generation-to-generation rather than being invented by each generation. Although nothing prohibits people from different places and times from arriving at exactly the same set of process steps, it is also reasonable to conclude that similarities may show ancient derivation from a common source. The convergence of need and resolution imprints itself in the material waste recovered by archaeology. From that admittedly incomplete record, archaeologists are challenged to peer into the past with a battery of analytical tools and fill in the rest of the story.

The concept of controls provides the lithic equivalent of DNA. Controls tell us what a craftsman understands to be essential for achieving a desired outcome. Uncontrolled actions introduce variation, and so cannot be as important. Therefore, every repetitive activity by early people can be expected to use some form of process control. Where the controls are placed and how stringent they are tells a great deal about the group. By comparing process recipes and controls, it should be possible to detect lineage of lithic technologies as people moved in place and time. Presumably, as a system was retooled to meet new needs, the most useful controls would be kept. Relationships between tool types should be measurable by how many process controls they share. We should be able to collect and process archaeological data within the same analytic framework that modern manufacturing uses to make production lines effective.

Dimensional specifications are usually used to describe tool typologies, but most analysts relay on a gut feel in assigning a type designation. I believe that inherent process controls that create the consistent flake scar patterning and overall feel an analyst instinctively applies. Tradition and ritual assure that the product remains unchanged long enough for types to be recognized in the archaeological record. Lax quality control, though, sometimes makes it difficult to separate legitimate type distinctions, especially when tools are at the end of their useful life.

The controls that allowed Folsom fluting were invented earlier by Clovis people, but needed to be refined before the ultimate expression of fluting could be routinely realized. Later styles of projectile points lowered the risk of failure by using simple process steps and controls that tolerated variation better. Through time, process steps and controls were also rearranged to accommodate changing priorities as people settled and widened their diet choices. Full appreciation of how processes were adapted from group to group will only emerge by deciphering each process in turn and comparing one to another.

I believe conscious decision making leads us on a different track than instinctive behavior. Underlying the decisions we make collectively is a conviction that "life can be made easier." That is somewhat different than "fitness." Houses are imagined to be better places to live than caves, and air conditioning is perceived to be more comfortable than sitting under a desert sun. Mankind has arrived at those notions through progressive accommodations over time, and we continue to chase an elusive paradise. As we explore the parameters of technological decisions faced by makers of fluted points, it will become increasingly clear that people had very good reasons for doing what they did.

Chapter 4

Basics of Fracture

In the previous chapter we saw that change is unavoidable, but how can we derive behavioral information from broken bits of rock? Early humans learned, through generations of experience, which hammers and strategies gave them useful results. This chapter may seem technical, but to avoid subjective opinions, it is necessary to raise flintknapping to a science. We will use the language of physics to clearly characterize how early craftsmen mastered the working of brittle stone.

Earliest flintknapping consisted of hitting a stone with a hammer, usually of stone or antler. Percussion is effective for removing large flakes to create heavy tools such as cleavers and handaxes, and the flakes themselves can be used for a multitude of tasks. Later, it was found that small flakes could be removed by pressing a bone or antler tool against an edge. The ability to use pressure to refine tools made by percussion allowed a wealth of subtle variations and refinements in tool forms, including projectile points and knives. [Details about the craft of flintknapping can be found in the book *Old Tools—New Eyes: A primal primer of flintknapping* by Bob Patten (1999).]

To quantify changes in lithic technology in a meaningful way requires a clear distinction between how the laws of nature guide fracture and what was imposed by the conscious design of man. While experimental flintknapping has been a time-honored part of archaeological reconstruction, it suffers from the occasional criticism that a modern craftsman cannot hope to know what actions were really taken by ancient workers of stone. Until recently, no sure means existed to associate a particular path of fracture with the conditions necessary for its formation. Today a modern tool, the computer, can be used to accurately represent what conditions lead to specific patterns of breakage in stone. The lessons of physics, backed up by experimentation and archaeological data, allow us to confidently correlate the means by which ancient workers manipulated their knapping tools to create complex tools out of stone.

Sometimes, in today's world of technological wonders, it seems that we can bend nature to our will. In reality, nature simply does what it does. Early native craftsmen were also bound by nature's laws, but had to learn to cope with them by trial and error. Physics provides a useful means to anticipate the result of an action without having to conduct thousands of trials to arrive at the same conclusion. We have learned to use mathematics to express the action of physics in a clear-cut way in order to describe the circumstances that will cause the outcome we desire. Having spent more than forty years learning to break rock the way I wanted it to break, I welcome the addition of theoretical modeling to our analytical repertoire. Selected experiments still have to be conducted to assure that the predictions work, but now they can be better structured and referenced to an overall body of knowledge.

Computer model

I was trained in Civil Engineering and was fortunate enough to meet a local oil engineer with a graduate degree in archaeology, Tony Baker, who took up the technical challenge of using computers to model fracture. We have enjoyed a fruitful collaboration, combining our technical background and intimate familiarity with Paleo-Indian artifacts to test each new idea.

Engineers have a way of reducing elaborate problems into the simplest elements possible. Once the simple element is understood, it is relatively easy to combine the elements and describe complex structures. Our basic model of a stone to be broken is a cube impacted by one force and supported by another. A finely gridded array of 5,000 cubes approximates a preform well enough for our purposes. A technique called *finite element analysis* identifies the point of maximum tensile stress, where a grid boundary is released to simulate a crack. Since each crack initiation requires a new solution, it takes a computer to perform thousands of calculations as the crack is extended.

Tony Baker (1998) used the ALGOL™ finite element analysis model to represent the actions of fracture. Because the ALGOL™ model does not include time, effects of velocity are masked. Tony enlisted the help of a physicist, Bill Watts, who used Mathematica™ to create a set of programs that describe the role of impact through time. The first program describes a rod-shaped impacter striking an elastic wall. Output is simply the time, after initial contact, before a crack can start, and the portion of the impacter that is compressed. Pre-crack force builds up linearly over time until the platform strength is overcome. A second program models the core as a fixed cantilever beam loaded by the conditions output from the first program. This represents both kinetic and potential energy through the duration of the fracture. It also gives the rate of change in velocity of the core at the end of the beam. Even though the finite element analysis is inherently a static model, the rate of crack opening can be related through Bill Watts' programs to track the progress of fracture in time.

Simple preform shapes help keep computations manageable and, when the developed crack is drawn, they also illustrate the principles most clearly. A pure rectangular preform was adopted because the initial scope of the project was to simulate Folsom fluting. It was a fortuitous choice that led to our understanding a

Computer model cell frame

The preform is viewed as a large number of discrete blocks that can be analyzed independently.

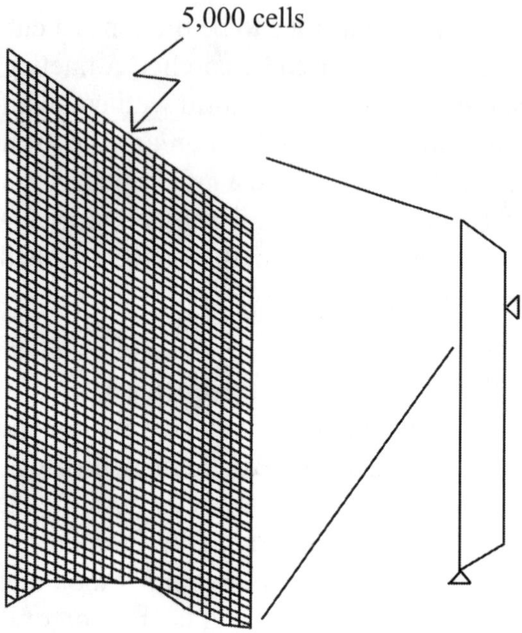

Force is represented by springs of a given stiffness

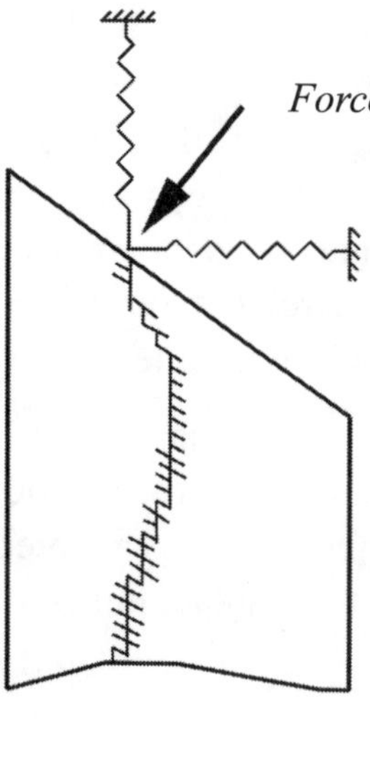

much wider set of problems than we could anticipate. Unlike real-world flintknapping, the results are not clouded by inadvertent changes in stone, hammer, or support conditions. Each time the same conditions are introduced, the crack appears the same. Equally important, each variable can be altered independently without changing the rest.

For convenience, Tony adopted an angle of 55 degrees between the striking surface and the core face, and it proved to be near optimal for fluting. No matter where we started the crack, within just a few cells it sought out the path it would have taken if it started from the point where the load was applied. This indicates that it does not matter whether the actual impact is spread out or concentrated, the prediction is still valid.

The computer makes it possible to chart the path of fracture for every angle of force, and can represent any kind of material. We described the action of each of the many fluting methods as force vectors, so fracture could be studied without worrying about how the initiating force was delivered. Although the mathematical model does not specify how force is conveyed, a number of approaches have been proposed for replicating fluted points, showing that one means of delivering force may be better than another. For example, exact angles of force can be determined by mechanical leverage. Other differences involve how fast or slow the load is applied, or how accurately it impacts. Many unique approaches to striking the platform were analyzed to explain their true differences in terms of physics.

Delivering force

As we have said, there are many ways to apply force to stone in order to make it break. Percussion was an early favorite, but later people came to appreciate the delicacy afforded by pressure. Whether force is delivered by impact or pressure, the motion of the tool dictates the direction of blow *and is important only just before the crack starts*. Platform angle, friction, or leverage usually alters the direction in which a blow delivers its force into the stone, so to be perfectly explicit about what happens at an instant of time, *angle of force* will be used to describe the mathematical vector.

The reference for angle of force is often chosen to be the longitudinal axis of the preform, so a blow in line with the axis is at zero degrees. While that allows a reasonable mental view, it happens to be mathematically inconvenient. In practice, the *real* zero angle of force can be physically described as the force that would only move the core forward, without turning it in any way—like the force needed to drive a nail straight. Moving the angle of force one way or the other from the true zero reference angle causes a torque that can be felt. Without support, the reference axis passes through the center of mass, but adding support moves the effective center of mass in the direction of the support. Although it is difficult to draw true angles of force, the craftsman has little trouble recognizing them and inclining the blow correctly.

The force of knapping develops two kinds of energy: kinetic and potential. Kinetic energy puts an object, be it a hammer or core, into motion. No fracture can be attributed directly to kinetic energy. However, deforming an elastic object that wants to retain its original shape stores potential energy that can be spent during the making of a crack, by vibration, or simply as rebound to the original shape.

If no crack forms, the impact causes two objects to bounce off each other. Since the mass of the two colliding objects obviously does not change as kinetic energy is exchanged, the hammer velocity must fall while the preform gains velocity. Anyone who has played pool will be familiar with how a moving object transfers its motion to the stationary ball. When the two objects are the same size, shape and mass, the relationship is equal but opposite. In other words, the collision of two object of unequal mass imparts the greatest speed to the lightest object. Too light a hammer would rebound without effect. Too large a hammer would just knock the core away.

Thinking of the hammer and core as springs helps to clarify how the blow should be delivered for optimum effect. Actually, the spring principle is of great use in many sports. Whether hitting a baseball, swinging a golf club, or using an throwing stick to fling a spear, each uses potential energy stored by elastic compression to propel something. Stone, iron, or antler, each material works best at its own unique frequency. Very stiff materials use potential energy faster than limber ones.

The ability to alter a hammer's stiffness is important because it provides a range of versatility to a knapper, without requiring him to change tools. A pressure tool gains stiffness as the tool is shortened. Percussion further increases the stiffness of the static condition by the speed and mass of the hammer. In our case, a percussion tool can be given a very large range of characteristic stiffness by swinging the hammer either slowly or fast. For example, a fast blow does not allow as much deformation as a slow blow and consequently stiffens both the hammer and the core. An inclined blow with a slender hammer also effectively lowers stiffness. Fracture is governed by how a blow is delivered, as well as by the structure of the hammer. A knapper uses elastic rebound of the hammer to speed the fracture much more effectively than if the hammer remains fully deformed while the fracture occurs. This helps us understand why forcing a blow with too much mass or muscle leads to poor flake formation. At the optimum, rebound breaks the contact between hammer and preform just as the flake reaches completion.

The energy a knapping tool shares with the preform can come from mass and velocity, or by using power from muscle or leverage. Either way, both the hammer and the core are deformed and posses potential energy. The greatest amount of potential energy is imparted to the object that is compressed the most—that is, the least stiff object. When pressure is used, the potential energy is stored primarily in the presser tool. Percussion, however, stores more energy in the core, and reaches higher levels of energy. So, a hammer made of rock develops more potential energy in the core than does an antler baton of the same mass. Whether the deflection is great or slight though, the hammer deformation is restored in the same amount of time. If we know how much deflection occurs, we can use this to estimate velocity.

For a given angle of blow, a flake can propagate within a range of initial potential energy. No matter how much energy is applied, the flake follows the same path. When there is no more potential energy, the crack stops. Since energy content of a blow is not readily seen or measured, the knapper must develop an intuitive sense of what hammer mass or velocity is appropriate for a given situation. Human physiology provides enough sensitivity through biofeedback and muscle memory to allow sufficient control of the blow for a knapper to optimize flaking conditions. Even though it requires muscle coordination to deliver just the right blow, there is still enough mechanical control in platform placement and angle of blow that a person of average talent can achieve predictable results.

Pre-crack time

Pre-crack time is the time needed for the hammer to transfer enough energy to start a crack. As long as the core can be loaded for the length of time it takes to complete one cycle of its natural period of vibration, we consider that to be static loading. Knapping by pressure always delivers static loading. Percussion usually produces dynamic loading because the load develops in less than one cycle, but it can also produce static loading. As we will learn later, dynamic loading can produce both the most useful flakes and the most undesirable.

Impact sites

I use point of impact to be specific about where on a platform the blow arrives. The *point of impact* is best described by how far it is offset from the face plane of the preform. If the offset is too slight, only a small, wedge flake can be produced. On the other hand, an overly large offset can force a flake deep into the body of the preform. For a knapper, the challenge of hitting such a precise spot seems daunting, but nibbling the corner of a square edge off with a short bevel can solve that problem by letting the knapper strike an edge but still have an offset. Computer simulation shows the same path of fracture whether the edge of the model has been reduced or not, and experiments demonstrate that as well. Archaeological flakes with carefully and deliberately reduced platform edges indicate that offsets were adjusted to control where the blow would land. Anyone who tries percussion knapping quickly learns the futility of achieving a precise impact anywhere but at the edge margin.

Flakes of any kind may be made over a range of offsets, but a flake that travels from one edge of the preform to the other has an optimum offset that accommodates the widest range of blow angles. Since platform offset can be precisely controlled by judging the distance from a straightedge placed against the preform face, it provides a convenient way to "stay in the groove." As we will see later, Folsom people took advantage of this principle to make remarkable, delicate projectile points.

Knowledge of the offset principle is evident even in the European Levallois technique, over 100,000 years ago. Instead of moving the impact site by relocating the margin, Levallois workers aimed at the intersection of flake scars that were deliberately placed to accommodate off-margin strikes with an appropriate offset. Results were not particularly elegant or standardized, but the fact that the concept was deliberately used represents a vast improvement over previous approaches.

It took Tony Baker and myself some time to recognize the significance of edge angle because the initial model geometry gave such good results. Eventually though, after trying various platform angles, the best results occurred at about 55 or 60 degrees for fluting. Since archaeological examples had not suggested this to be an important variable, we reviewed beveling by other knapping traditions and found that they naturally develop 55 to 60 degree angles. In fact, conscious effort is required to force the platform angle to a different value. It is not clear whether the people who practiced fluting were particularly sensitive to achieving an optimum platform angle or if they just used the convenient control that nature provided.

Platform strength is perhaps the most important and most subtle variable in knapping. The knapper has to use experience with a given stone to anticipate failure at just the right level of energy. Since platform strength sets a limit on how much potential energy is available to make the flake, the platform must be strong enough for the flake to reach as far as the knapper wants it to. A

Edge battering

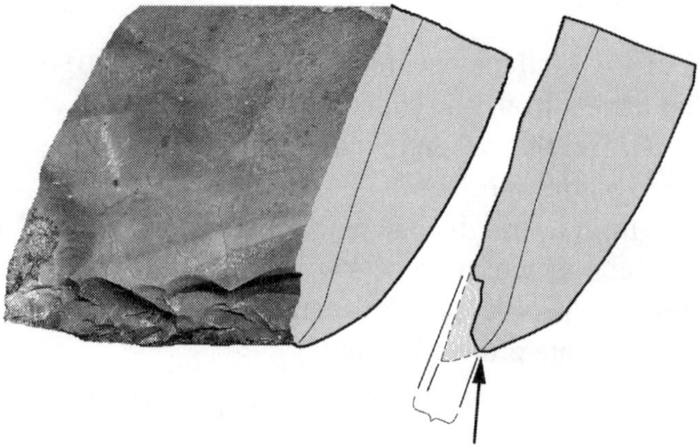

Reducing the edge margin by battering allows a knapper to quickly obtain a suitable offset to make a flake of the requisite thickness. Careful beveling achieves the same result, but requires more work. Without the offset, only thin flakes can be made.

platform that is too weak allows crushing; too strong and the flake cannot complete without a very energetic blow. When a thicker offset is required to create a suitably thick flake, it also can make the platform stronger unless the margin is reduced. Once the appropriate platform is settled on, more subtle adjustments can be made by altering the blow and the preform support. Observing whether a tradition typically built their platforms strong or weak helps in identifying the preferred kind of hammer material. For example, because the stiffness of stone or copper overcomes platforms more readily than antler, the sturdiest of platforms are required.

Starting the crack

For knapping to accomplish its end, the applied force must overcome the strength of the core and start a crack. The slow build-up of force by pressure stores the maximum amount of energy to be spent in fracture. However, it takes the mechanical advantage of leverage to reach levels of force comparable to percussion. Once the crack starts, potential energy is responsible for the flake's advance.

Only the component of applied force that is parallel to the platform face is responsible for starting the crack. When the crack opens, the flake behaves like a leaf spring that is as long as the crack has advanced. As a result, the impacter shares energy first with the core and then with a leaf spring of ever increasing length. The outward velocity of the flake platform as the crack extends matches the velocity of the hammer.

Although the mode of applying load to the platforms is usually characterized as either percussion or pressure, physics require us to make a subtly different distinction. Pressure flaking applies force slowly in the static mode until the platform strength is exceeded. In the *static loading mode*, where the loading time is slower than the natural period of vibration of the core, the core rebounds away from the flake. If an impact starts the crack before that time, however, inertia causes the core and flake to move in the same direction for a short time while the crack is being formed. This condition may be called *dynamic loading*. Because the crack is already on its way when the core finally rebounds, it can grow to nearly twice the length of a crack made in the static mode.

When a knapper works in the dynamic mode, there is a good chance he will use too stiff a hammer and create the condition for hinge terminations. When the crack starts within 0.15 of the core's vibration period, the progress of the flake will be interrupted by a hinge when the core rebounds. Slowing the blow to avoid the hinge threshold allows the crack to reach its full potential length without hinging. When load is applied faster than 0.085 of the core's period, the fracture is forced to the surface in the shape of a clam shell. This is the shape that inspired the term *conchoidal fracture*, but it represents the worst condition for knapping.

Hertzian Cones

Although the phenomenon of Hertzian cones has frequently been used as a way of explaining fracture, it can explain only limited fracture behavior. Severe conical compression under a straight-on blow tears a cone away from the surrounding stone. Hertzian cones depend on a hammer of low mass delivered at high velocity, which means that the hammer is in contact with the core for only a very short time, separating Hertzian cone formation from other normal means of flake creation. Because the hammer is relatively light, the cone can penetrate only a short distance into the core before the initiating energy is used up. Cones are limited to three times as deep as the diameter of contact with the hammer (Speth 1972: 38). The small diameter circle of stress associated with the very beginning of hard hammer impact is probably a form of Hertzian cone, but the remainder of the flake propagates by bending away from the core face.

Extending the crack

From computer modeling, we know that the path of an emerging crack is completely guided by the expansive rebound of the hammer, and core rebound will be spent in vibration. Regardless of whether the impacter has deformed from a percussion stroke or by pressure, it wants to rebound in the same time, as long as it is made of the same material. That is because the impacter acts as a spring and restores its uncompressed shape as a function of its natural frequency. As soon as a crack is initiated, the knapping tool tries to regain its undeformed shape. The rebound of the tool then interacts with the developing flake only at the impact site, and immediately becomes the dominant factor in how the flake detaches.

The strength of atomic bonds within the core sets limits on how fast a flake can progress. That quality is commonly recognized by describing a stone as being glassy when it breaks readily, or tough when it has a low characteristic velocity of fracture. The material characteristics of the hammer then determine how much time a crack can travel before the core brings it to a halt. Because the hammer separates the flake platform from the core, the flake can only travel while it is impacted. In the static mode of loading, as a crack lengthens quickly, bending stress of the flake lowers. Think of a zipper that takes a tug to get started, but runs easily until it encounters a thread in its tracks. This should manifest as a flake that is relatively ripple-free near its termination.

You might expect contact time to vary with the velocity of a blow, but that is not the case. The nature of freely colliding objects is such that they remain in contact for the same amount of time regardless of their relative size or speed. No matter how the core or hammer are constrained at the start of the blow, as soon as the crack starts, it uses so little energy that hammer rebound is almost instantly independent of any other parameter. The hammer will keep pushing the flake for half of one cycle of rebound. Because contact between hammer and flake relies solely on hammer rebound after the crack starts, the speed of the hammer has no effect on contact time. When a flake is removed by percussion, the hammer is usually in contact for less than five ten-thousandths of a second. Obviously that does not give a craftsman time enough to react during the blow, and any correction is impossible once the crack starts.

The time available for a crack to occur is controlled by hammer compression, unrelated to the speed of fracture. Fracture speed is primarily a function of the length and flexibility of the core. Swinging the hammer faster can produce longer flakes because the fracture starts in the dynamic loading mode and finishes in the static rebound mode. The spring effect keeps the contact between the hammer and the core constant. The more flexure that is stored in the hammer prior to the crack starting, the further the flake can travel during the contact.

In our perfectly simple computer model, flake paths are independent of the level of energy applied prior to the start of a crack, which would seem to assure that the speed of propagation should make no difference. However, slow fractures are heavily influenced by variations in the structure of the stone, while very fast flakes cut through different materials with equal facility. The difference is visible in the texture of the fracture surface, with relatively slow pressure flakes exhibiting greater roughness than percussion flakes from the same stone. Of course, real stone is much more complicated than theoretical stone. It is important to keep in mind that once energy is spent, the crack must stop, even if it is on the correct path. In that case, the flake terminates as a step.

Surface topography

From the outset, the front face of the computerized preform model paralleled the back. Initially, it appeared that such uncomplicated geometry was necessary to insure that the preform stiffness remained constant as the fracture progressed. Adding thick areas to the face of the preform model causes the fracture to rise when extra stiffness is encountered or lower when a thin spot decreases stiffness. Experiments indeed confirm that surface geometry variations steer the fracture path, especially when pressure is used to remove the flakes. Well-defined central ridges, like those of Cumberland-style fluted points tend to exhibit ripple patterns of small radii because the center of the channel flake is relatively stiff. Less domed preform ridges result in ripple arcs of relatively greater radius. And when a second flute is made without re-chipping the surface, the double ridge of the channel flake surface reflects itself as a W-shaped ripple scar when viewed from the tip.

The shape of the core surface provides directional stability of a crack. The thickness of a dorsal ridge is often enough to keep the fracture plane on course but where the dorsal ridge goes, so does the course of the flake. Biface flakes that extend past the centerline of the preform often show this effect where the flake outline follows the intersection of flake scars, otherwise known as arrises, well away from the starting axis of blow.

When thickness is added to the back of the preform, however, there is little effect on the flake path. Archaeological observations that fluted preforms typically have flat backs therefore require an explanation that has nothing to do with preform stiffness. Experience shows that adjustments to surface topography are far easier and less risky when the needed adjustment is relatively slight, and that is especially true for thin preforms. By flattening the back face of a preform prior to the first flute, the makers simply lowered the variance of surface that had to be overcome to provide a straight face profile for the second flute. The principle of starting with a flat preform blank was also useful to other flaking strategies, and persisted long after fluting fell from favor.

Effect of surface on crack path (over twice actual size)

Each of these flute sections has the exterior preform surface to the right and the flake edge is traced at the center of each profile.

A. Irregularities in the preform cause the flake to thicken as surface bumps stiffen the flake, and thin when depressions are encountered.

B. A straight preform crest is needed to guide a straight flute scar.

C. If the preform is not flat, the flake follows the surface contour.

D. Greater thickness of a channel flake lessens the effect of surface irregularities on the fracture path.

Support

Support accomplishes more than keeping a core in place during percussion—it is like adding mass to the core. Very large cores have enough inertial mass for flakes to be broken away without any external support. Eventually, enough flakes are removed that support must be provided to continue flaking. By adding just enough support to make flaking possible, the knapper can make the longest flakes since they have the longest time to progress. Effectively, adding support increases the stiffness of the hammer. Support often dominates stiffness, regardless of the relative size of hammer and core. Because the hammer should be less stiff than the core, too much support can be detrimental to flake formation. Subtle variation in the degree and placement of support allows a knapper to manipulate the preform for optimal control of the fracture. Most of the following discussion is designed to explain fluting, but it also applies to how support operates during bifacial flaking.

Support acts on the core in the same manner as the hammer. The effect is the same as impacting both ends at once. Newton's Cradle, where three steel balls are suspended by strings, illustrates the condition when the middle ball is considered as the core, the one on the end as the support, and the one swung against the middle ball as the hammer. The stress on the core depends on combining the respective hammer mass, stiffness of the hammer, and the support.

Extra support seems to be reasonable when a platform is just too strong, but it can lead to some very undesirable consequences. From modeling, we know that overly stiff hammers or too rigid support of the core can promote hinge fractures. As a core or biface becomes smaller with the removal of each flake, its natural period of vibration is shorter and hinge terminations will inevitably limit the flake length. That effect limited most prehistoric use of stone or copper hard hammers to relatively coarse work at a quarry, while refined flaking was done with soft hammers. Users of copper or stone hard hammers sometimes plan for short enough flakes that the hinge condition is not reached before the flake completes.

Newton's Cradle

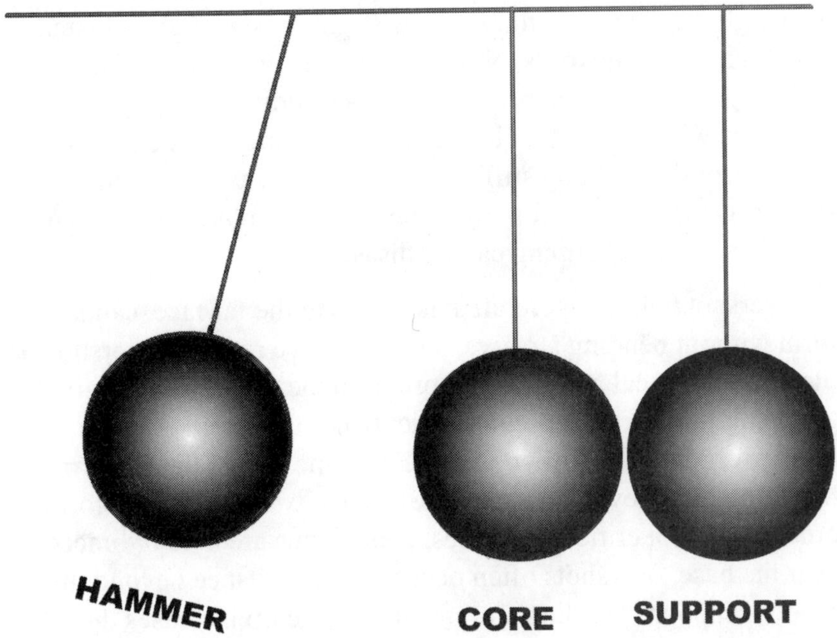

Newton's Cradle shows how increasing support stores more compression in the core for rebound.

Lateral support location

For the past fifty years, experimental fluting depended on supporting the preform either by clamping the edges or using a solid support near the base. That unrecognized historical bias was reflected in our computerized model for the first year of investigation, but both the model and experiments showed an unfortunate tendency toward early overshots. Because overshots were so clearly influenced by bending at the upper support, the model was eventually rerun with the lateral support moved to a half-inch from the tip. Surprisingly, the setup worked much better than expected. Even where an overshot was obtained, enough length remained to allow further fluting. An unexpected result was that near-tip support made a more direct angle of blow feasible and lowered the opportunity for bending. Interestingly, flake thickness increased slightly when the lower support was introduced. Although the change in thickness is slight, it is enough to match aboriginal flute thickness. Now we can see surface irregularities, and support high on the preform edge, as adding bending that interferes with a straight, clean flake. Clamping the preform edges can be useful by adding stiffness to a preform, as long as the support is perfectly uniform and the flute is removed by pressure. Otherwise, edge clamping can be disastrous.

Overshot flakes, where the flake exits to the far face, cannot form without bending stresses, so the relative reach of overshots in the archaeological record is important to the question of original knapping practice. Vises can restrict bending so well that flakes rarely overshoot, and often exit the extreme end of the preform tip. Compound leverage techniques typically clamp the preform in order to control critical variables. If preforms are held or braced near the base, overshots often occur a short distance beyond the point of support. Holding a preform near the tip increases the likelihood that overshots will occur near the tip. Low lateral support is difficult to use with any technique other than percussion.

When the concept of support location is applied to flakes used to make bifaces, it explains how Clovis knappers could control overshots. By adjusting how a preform is supported, it is possible to deliberately cause one flake to travel straight and flat by constrained support about the preform perimeter, and make the next flake overshoot by holding only the far edge. Once the means of controlling flake curvature through adroit manipulation of support is mastered, an artifact can be contoured at will. Combined with strategic impact spacing and sequencing, it is possible to achieve percussion flake scar patterns as striking as those normally seen only on pressure-flaked tools.

Because overshot flaking is often cited as a signature of fluting technology, it is important to note that experimental knappers are most familiar with unintentional, often catastrophic, overshots resulting from improper support. Overshots produced by Clovis and Solutrian knappers seldom removed much of the far edge.

Options for lateral support location

A. Upper support allows various blow angles, but increases risk from overshot.

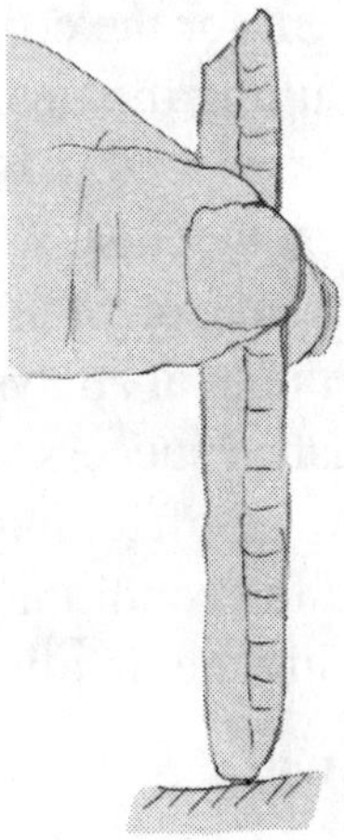

B. Lower support is less risky, but the preform must be struck at the correct angle.

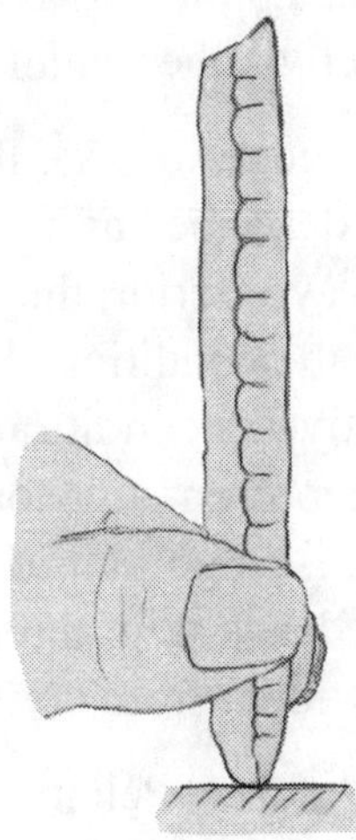

It is more appropriate to focus on the intent to extend the flake trajectory from edge-to-edge, rather than on the occasional error when the flake traveled a bit too far.

Preform-tip contact

To begin with, there was no apparent reason for any special requirement for the contact between the preform tip and the anvil, beyond dulling the tip by grinding. Some Folsom sites, though, show that tips tend to be beveled toward the face that was fluted last. To test the idea, we ran several simulations with all variables held constant except for where the tip contacted the supporting anvil. In the final analysis, it was evident that fluting could be accomplished successfully without extra effort to position the tip. However, offsetting the tip support the same as the platform expanded the conditions over which full-length flutes could occur. Re-chipping the tip after each flute adds labor but lessens the risk.

Flake extent

The computer model was designed to represent flakes that progress edge-to-edge across a core or preform, but it applies to every other kind of flake as well, from short, feathered flakes, to step flakes, and overshots. Furthermore, while the model was initially presented as the longitudinal section of a fluted projectile point, it is just as applicable as the cross-section of a biface. By altering our perspective, the model applies to any lithic technology.

If an edge-to-edge flake is our standard, then any other flake can be described as a fraction of the standard. Fractional flakes are in no way inferior, they just serve a different purpose. From the model, the conditions leading to fractional flakes are as distinct and instructive as conditions for edge-to-edge flakes. Since fractional flakes represent any condition that does not allow an edge-to-edge flake, it is important to note whether the flake feathered, overshot, or ended as a step. Intent is much more difficult to assess, but is often evident from context.

Dr. Andrew Pelcin (1996) demonstrated through experiments that flakes made by impact achieve a mass that can be predicted by how far from the edge the load is placed. Flakes from flat

Positioning preform tip contact

Optimum results occur when tip contact is offset the same as the platform offset.

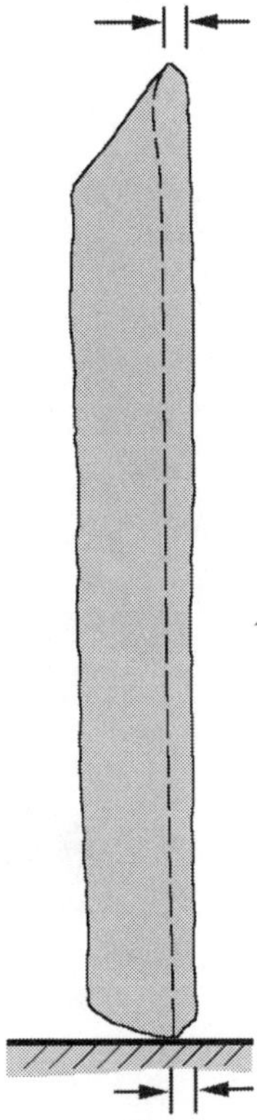

slabs spread into circular shapes, while flakes with ridges on their surface elongate in proportion to how distinctly the ridge is isolated from the surrounding surface. A raised surface imparts stiffness to a flake, which determines its ability to extend. Therefore, a flat-surfaced flake is forced to expand greatly to acquire the same stiffness that a narrowly ridged flake has.

By mapping conditions where various categories of flakes may be expected, we can see that edge-to-edge flakes are possible within a specific range of conditions of blow angle and load location. Impact allows fracture only until the hammer pulls away from contact with the platform. Since hammer velocity sets fracture velocity by how much flexure is stored, faster blows limit

Flake types

Full length

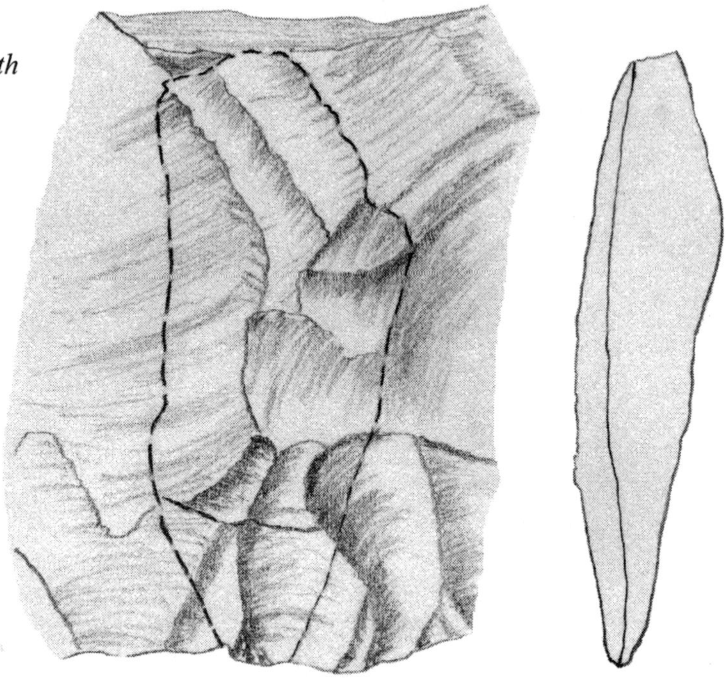

Overshot

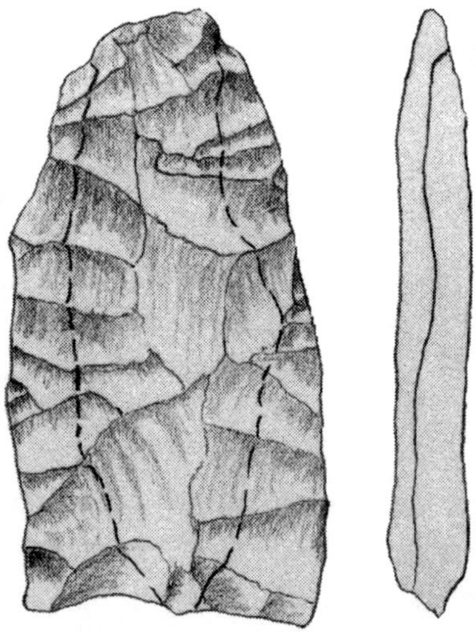

Termination types

Feather

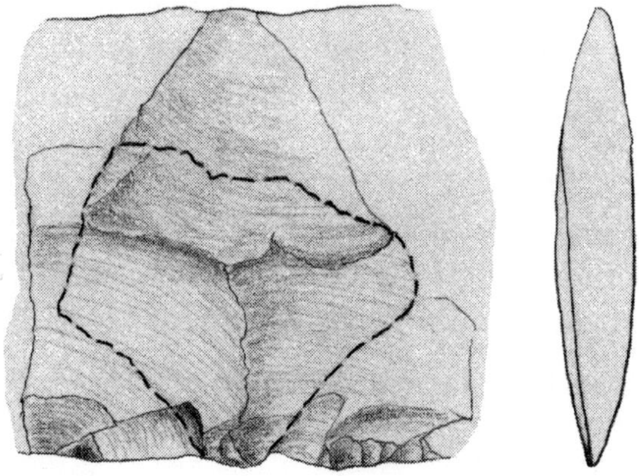

Step

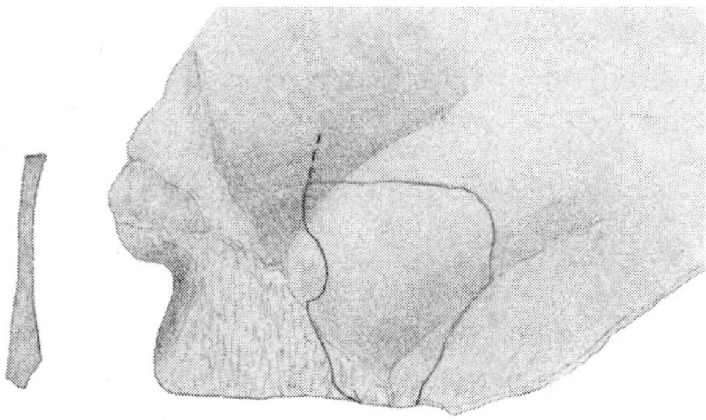

Hinge

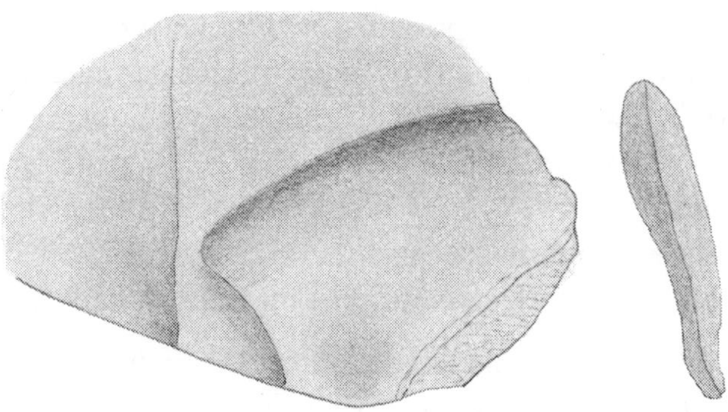

how far the crack can run. If the flake path is too long, it must stop short as a step termination when energy is used up. Static pressure can store enough compression that rebound of the knapping tool provides sufficient follow-through for the crack to develop fully. Feathered terminations represent the highest-angle blows, and overshot flakes occur when blow angles drop too low. Excessively short loading times, because of stiff hammers or support, cause hinge terminations and limit the flake length severely. In order to achieve the most impressive results, the knapper is forced to work between the conditions that limit flake length.

Choosing a knapping tool

Percussion tools

Matching core stiffness to tool stiffness requires judicious selection of tools and adroit manipulation. A knapper may have access to a wide variety of tools, but just as likely, he must make do with what is at hand. Provided the knapper knows how to properly manipulate the tool, he can then create nearly any kind of flake he wants.

While tool weight is intuitively important to impact, it plays less of a role than one might expect. Tools can vary considerably in weight because density makes more of a difference in the ease of handling a tool than it does in delivering energy. Antler combines a light, porous core with strong, dense walls to provide a strong, but easily wielded percussion tool capable of storing flexure energy.

Theoretically the direction of blow can vary through nearly a sixty degree range (Coterell and Kamminga 1987). Practically, the traction a hammer provides determines whether a knapper has to worry about the angle of impact. Hard hammers typically skid, so they work best when directed perpendicular to the platform surface. Knappers can and do change the platform angle, but that is not always easy or convenient. Antler batons envelope the edge slightly, and so accommodate a wide range of impact angles without special adjustment of the platform angle.

For any given hammer material, spherical hammers are stiffer than slender ones. More importantly, in sharp contrast to a slender baton, the delivery of a spherical hammer cannot change its stiffness. For a slender baton of any material, jabbing within about 20 degrees of the long axis allows maximum stiffness and contact time between the baton and the core. Increased inclination allows the baton to bend, and damping keeps some energy from being available for fracture. Too much inclination eventually puts enough "waggle" into the baton that the vibrations can periodically lift the baton from contact. That release of contact wastes energy through vibration and robs a fracture of forward impulse. Judicious manipulation of a baton satisfies a wide range of stiffness and thereby provides the greatest of operational flexibility.

How fast a blow arrives at the platform can influence the tool stiffness and the energy available to overcome platform strength. In turn, the mass and shape of the tool can aid or hinder how fast the craftsman can deliver a blow. It is especially difficult to reach high speeds with a short jabbing stroke. Arced blows easily achieve high speeds, and can be very accurate if a small radius of rotation is used. Weight and bulk clearly interfere with a speedy blow. Fortunately, the heaviest tools work best when the blow is delivered relatively slowly.

Choices have consequences, so ultimately, the serviceability of a percussion tool is due to how all the variables are combined. Any specific level of kinetic energy can be attained by nearly any kind of hammer by adjusting weight, stiffness and speed. The canny craftsman prefers a tool set that requires the least adjusting. For mobile people, carrying a graduated set of hammer sizes in order to accomplish a range of tasks would be cumbersome. Antler accommodates work on the widest range of preform shapes, through the greatest range of angles of blow. The same antler hammer can be used to remove flakes from a compact core as well as to refine a delicate biface. It is also lightweight and easy to carry. Copper hammers are not known archaeologically until Archaic times, but behave much like stone hammers, although they are even heavier.

Pressure tools

Pressure can duplicate a percussion flake, without changing the angle of blow, only when platform strength is increased. Too weak a platform allows the crack to start before sufficient potential energy has accumulated to send the crack far enough. Consequently, leverage fluting by pressure typically requires squared and heavily ground platforms, while percussion fluting of the same preform can be generated with a much more delicate platform. The slow loading rates of pressure insure that hinge terminations cannot occur, yet overshot, feather, and step terminations are still possible.

Modern knappers utilize a wide range of pressure tools, but almost any tool works well if it allows flexing to store energy. If the pressure is added by a rigid knapping tool with no flexure, then the crack will use the applied energy very quickly and leave a step termination. Increasing the flexibility of the tool increases the follow-through, and provides enough contact time with the tool to drive a flake to completion. Part of the flexure can be in the tension of a knappers arm. Since copper does not store enough energy in deformation, more work is required to make an acceptable flake. Also, copper's stiffness can shatter a delicate platform more readily than an antler tool would, so knappers using copper pressure tools need to grind platforms more thoroughly than antler pressers require.

Antler, horn, and ivory all perform well as pressure tools, but do not come in equally convenient starting shapes. Antler tines are most readily put to use with a minimum of effort. Horn must be carved and straightened, and even then the walls are often unreasonably thin. Ivory takes even more work to fashion and requires reinforcement to counter bending. Copper can be used, but it is as least twice as stiff as the organic options. The most suitable level of springiness and strength can be achieved by fitting any of these materials with a wooden grip.

Ishi stick

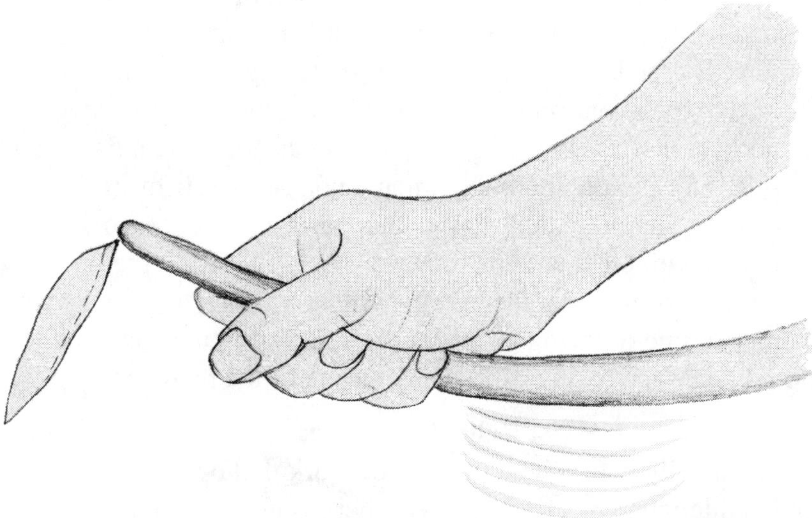

An "Ishi" stick stores energy in flexure of the stick, as well as the knapper's muscle and tendons. Ishi was the last known Indian in California to live without the constraints of white civilization.

Applying the principles

It is useful for the craftsman to know how to anticipate conditions to make a specific flake because removing a flake alters the behavior of the core, and a biface will not react the same way for two identical blows that remove flakes. Considering the difficulty in matching variables exactly, it may seem next to impossible. The solution rests with the knapper's "factory-installed on-board computer." Experience and biofeedback provide reliable guidance even when the knapper is not overtly aware of what is going on. However, conscious understanding and deliberate adjustments greatly accelerate the process of tuning in to the right set of conditions.

Like any skilled athlete, a competent knapper can actually feel the difference between an effective action and one that doesn't work. While the interaction between variables can be described in academic detail, the craftsman need only do what feels right.

Since it is not usually feasible to make continual adjustments to mass, varying angles of blow and inclination of the tools provide reasonable means of manipulating the stiffness that controls fracture. Biofeedback gives the knapper enough information to adjust tool mass and preform support to achieve effective levels of energy transfer. Even a light tap that does not make a flake can provide information about alignment and relative stiffness. Every stroke of the hammer adds to the knapper's experience, so after a bifacing session it should be relatively easy to deliver the fluting stroke without fear that the energy will be mismatched. Likewise, after a lifetime of knapping under the same system, a knapper should be effective on nearly every stroke of the hammer.

In describing ancient technology, we can assume the knapper used a socially determined manner of work that operated within a predictable range of effectiveness. There is no need to quantify the load placed on the preform because it is taken care of automatically by conditioned response of the knapper. The path that the fracture will take, however, has more to do with the impact site, surface morphology, and support placement. Those variables can be observed, measured, consciously manipulated, and simulated by the computer.

Experimental and archaeological correlation

For fracture mechanics to be accepted as a tool of archaeology, it must be validated by experimentation and hard data. A three-way interaction that plays theory, data, and practice in a dynamic tension of constant questioning cannot fail to yield valuable insights. The observations that follow are intended to spur as much curiosity as they satisfy. Do not expect all the mysteries of stoneworking to be laid to rest here—this is only a beginning.

By dropping steel balls on plate glass, Dr. Andrew Pelcin (1996) discovered relationships correlating blow angle and impact placement with the resulting flake mass, which led directly to Tony Baker's (2000) computer modeling. Pelcin's model helped Baker predict satisfactory results, and it also helped calibrate stiffness constants that explain fractures in glass.

When Pelcin dropped steel balls of varying mass onto the edges of plate glass, plots of flake lengths defined a neat, linear relationship, for the most part, but deviated mysteriously when the mass reached a critical threshold. The deviation indicated that critical independent variables were unrecognized or uncontrolled. Now we know that the added mass altered the relationship of hammer stiffness to core stiffness, and hinge flakes appeared when the natural vibrational period of the core was at least 6.7 times longer than the time taken to start a crack. Increasing the hammer mass further only made the problem worse. Fortunately, the value of the data was ultimately recognized, once there was a theory to make sense of it.

Pelcin experimental setup

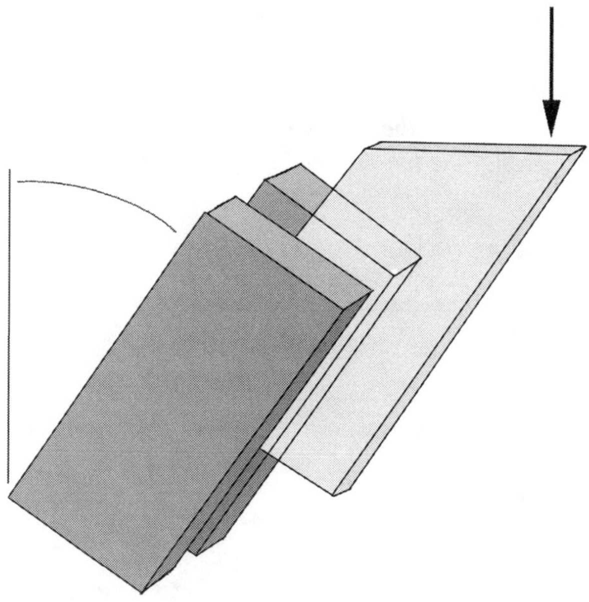

A steel ball dropped on the edge of a plate of glass does not represent normal knapping because the clamp holding the glass creates too much support. However, it does allow reliable experimental data to be controlled by specified variables.

Flake data

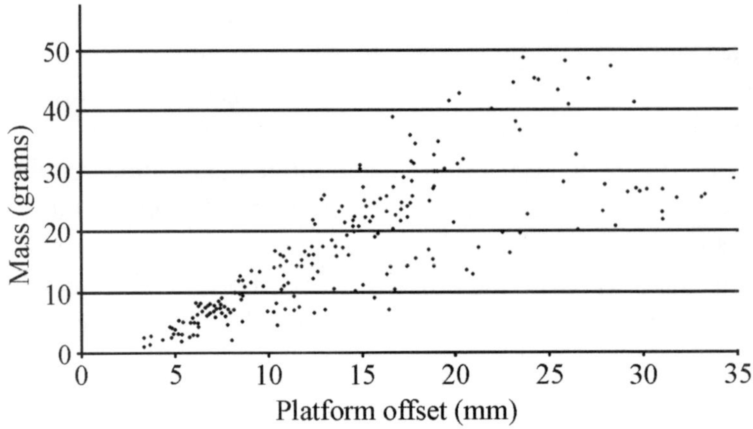

Dr. Pelcin (1996:178) found a consistent relationship between flake mass and platform offset. Some data, however, branched in an unexplained trend. Later computer modeling proved that the deviation occured when the core's vibrational period exceeded 6.7 times the time needed to start a crack.

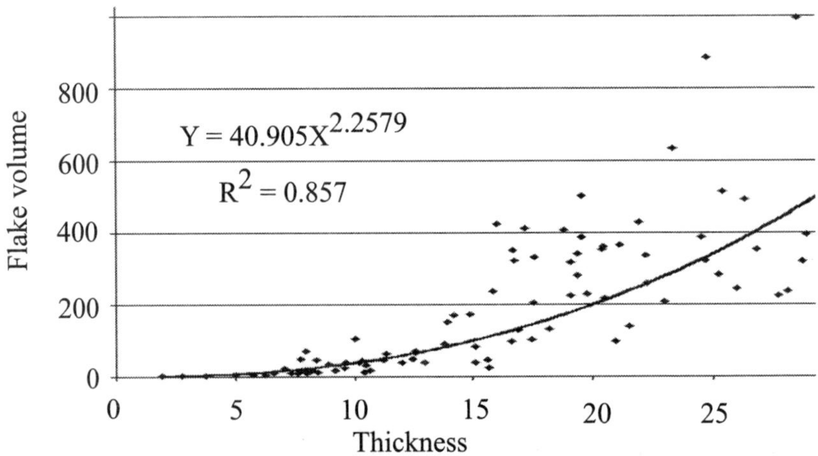

A similar plot of flake volume against thickness shows that the data fits a power law. For a given thickness, volume increases with the area of a flake surface.

In theory, flutes from a thick preform could be expected to average slightly thicker than those taken from a thin preform. An analysis of flutes made by Folsom people, however, shows no detectable difference in thickness between first and second flutes. The best explanation appears to be that platform offsets were made to a dimension that best fit the overlap between possible channel flake thickness for first and second flutes. By controlling platform offsets, makers of Folsom points were able to use the same process for each face of a preform and expect equivalent results. Even though it is theoretically possible to make ad hoc adjustments for different phases of the fluting process, it is easier to choose an operational standard in the mid-range. Early people were probably not able to "dial-in" precise angles of blow and stiffness, but it is likely they understood the relative effects. Subtle adjustments from one stoke to the next allow just enough extra control over the depth of channel flakes to assure a reliable, and thus standard, outcome.

Once appropriate flakes were known to be theoretically possible by direct percussion, modern replication was required to prove it a viable means for fluting. The highly dulled tips of archaeological Folsom preforms were imitated, to protect the tip as it was driven against an anvil. At first, the results were encouraging in that long, clean flakes could be removed. But, it was extraordinarily difficult to get consistent results because the aim-point was seldom in the same location for consecutive blows. That problem was solved by positioning the anvil to place the fluting platform in exactly the same location and orientation required for bifacing. Using the same kind of stroke to remove channel flakes or reduce bifaces imparts enough control to make direct percussion fluting reliable.

Knapping experiments quickly proved the feasibility of using percussion for fluting, but they also disclosed an unfortunate trend for overshots near the preform base. The problem was traced to bending stress introduced by the fingertip support. While moving the support nearer the tip initially seemed counter-intuitive, the computer model showed that it should greatly reduce overshots even though the angle of blow needed to be more precise. Overshots could not be entirely avoided, but they could be moved close to the tip where enough length remained to complete a serviceable projectile. My continued experiments suggest that allowing the percussion blow to drag the preform loose from the fingertip support constitutes a control on angle of blow because it trains the knapper for the highest effective angle of blow, where overshot is minimal.

The requirement to maintain a balance between hammer and core implies that knappers should not add power to get flakes to travel farther. It is hard to accept that minimal effort should be expended while doing percussion knapping, without experimental confirmation. Using a hammer of fixed mass to reduce a biface eventually sheds enough core mass as debitage that flakes fail to propagate effectively. Faster, or more powerful blows do not

Angle of blow vs offset (2” preform with upper support)

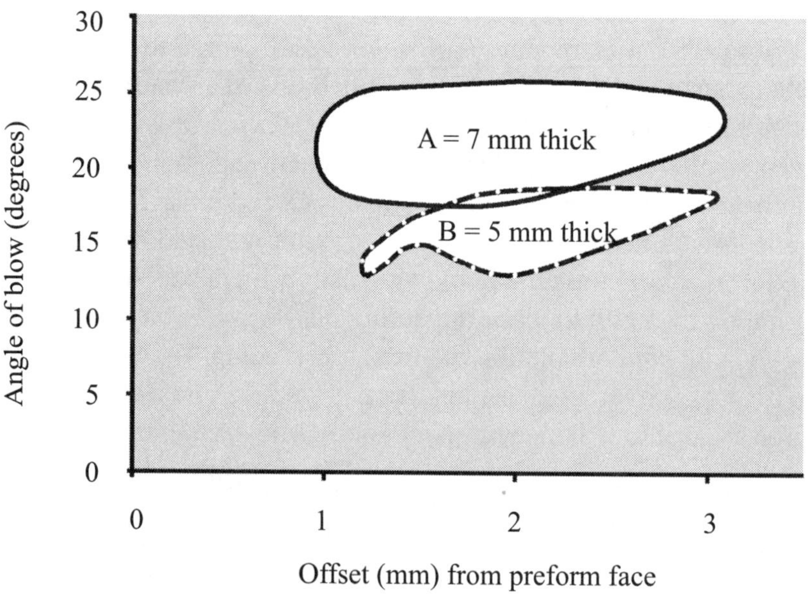

This diagram shows the window of opportunity to make full-length flakes on a specific preform. Tip support both lowers and narrows the usable range of blow angles, as well as allowing a slightly thicker flake. Higher angles of blow produce feathered flakes, while lower blow angles cause overshot flakes.

help. At that point, only fractional flakes are possible and step terminations begin to predominate, as predicted by the model. I found that sufficient mass can be restored by the simple means of pressing a stone against the reverse face of the preform. Once enough mass is added, the same hammer can again be used to remove flakes effectively. Under this principle, it is a simple matter to add mass so that ultra-thin bifaces can be flaked to completion without supplemental support. Projectile preforms are so light that they require anvil contact in order to increase effective hammer stiffness sufficiently for fluting.

During one demonstration, in order to accommodate a particularly difficult quality of stone, I reverted to an old practice of knapping without using a pad to receive the blow. Since the results improved even more than I expected, it forced me to re-evaluate which controls were active. It turned out that I had been focusing on the impact of the hammer with the pad and missing the biofeedback from the impact with the stone. By directing my attention to the first impact, I was able to return to effective knapping no matter what setup I used. Two lessons can be taken from this experience. First, biofeedback is effective if the knapper is attentive to the right cues. Second, it highlights the importance of maintaining proper relationships between hammer and preform. For a workman accustomed to reacting to the stimuli of nature, the behavior would be automatic. For most modern knappers, it means we must be alert to what really matters.

Timing the crack

For glassy materials like obsidian, a fracture phenomenon called "*gull wings*" can be used to obtain an instant measure of velocity at a point along the fracture path. Gull wings are caused when the compression wave hits an imperfection and sends a concentric sonar-like ping that interacts with the fracture front. The angular spread between the wings can be converted mathematically to quantify the fracture speed (Tsirk 1986). Data from a number of knapping experiments show crack velocities starting at about 1,100 meters per second and slowing as the crack extends. The speed at which a fracture progresses is related to the flexibility of the impacter and limited by the strength of the core. Flake speed comes from rebound of the impacter (even with pressure) and depends on how much impacter deformation is present as well as the rate of rebound. One researcher, Karl Hutchings (1997), set out to confirm early atlatl use by measuring fracture velocities revealed by gull wing marks on obsidian. In the course of verifying that projectiles were broken by high-speed impact, he accumulated a data set that confirms the difference in fracture velocities between pressure and percussion. Generally, fractures made by pressure propagate at between 100 and 500 meters per second, while dynamically induced fractures develop speeds of 350 to 1,500 meters per second.

Summary

Although the basis of fracture seems very technical, for a skilled craftsman intuition plays a stronger role than reason. Our senses gather useful feedback that can guide us to the right actions because the same properties of rock that allow it to be broken determine how an impact sounds. A light tap of the hammer is often sufficient to learn how the hammer and preform will interact. We can shortcut the learning process and become "instant" experts by using science, but ancient flintknappers had to learn about the constraints of physics through generations of experience. Even armed with scientific knowledge, a modern knapper needs the training of experience to become proficient.

The literature is replete with suggestions that the mobile lifestyle of bison hunters demanded a simple and reliable tool production repertoire, and direct percussion allows the most portable of tool kits. More important were the skill and training required to be able to make the preform in the first place. The dexterity required to hit the same spot on a pad over and over is not especially difficult. Repeated strokes ingrain muscle memory of the worker for a task that was simple to begin with. Judging where each blow should be placed is also primed by the experience of what works and what does not. It is tempting to see makers of Folsom points as masters of skillful improvisation as they overcame new challenges of stone quality and form. Another

Gull wing formation

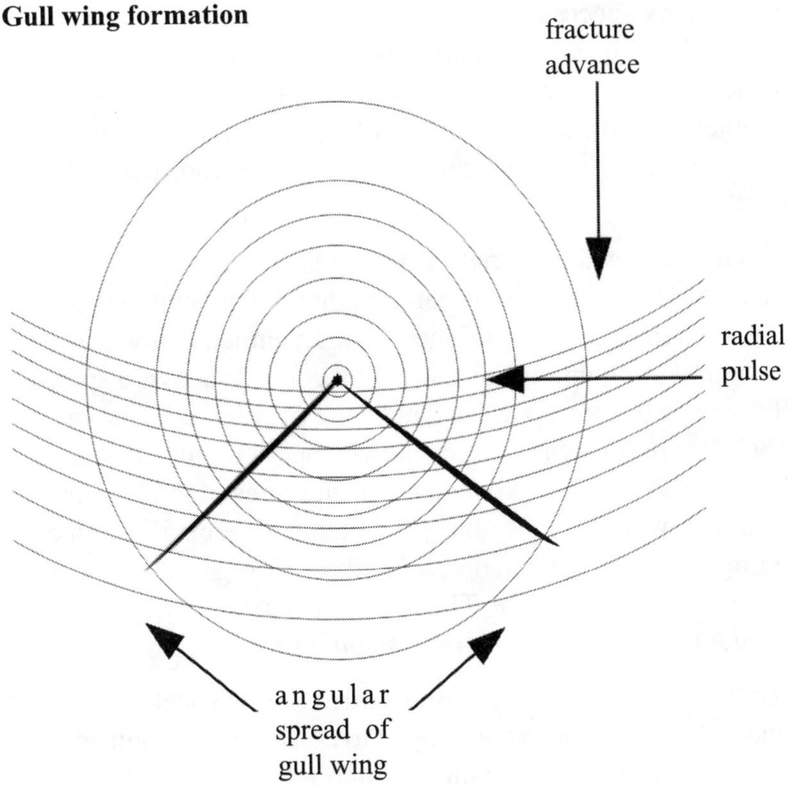

Intersection of advancing fracture with a radial pulse occurs when the compression wave strikes an imperfection and allows fracture speed to be directly measured by angular spread of the gull-wing mark.

possibility is that Folsom people were technically adept without requiring any extraordinary ability. For the group to flourish, everyone, no matter how awkward or adroit, had to be able to accomplish the same end equally well.

At times, early lithic technologists seem as competent as any modern engineer, but important differences should be acknowledged. While engineers see themselves in the position of harnessing the laws of nature in order to control it, early stone workers merely accommodated nature. By recognizing what influences steer fracture in predictable ways, knapping actions can be selected that lead naturally to the desired result. Constant improvement by successive generations, as each new variable allowed results to be altered, naturally evolved as a tradition of lithic technology. Near the same time that fluting reached its zenith, other non-lithic technologies, like basketry, must have taken increasingly important roles, and may have contributed to the decline of fluting.

When technological diversity is low, it is not unreasonable to use a complex set of process controls that are understood and mastered by a majority of band members. Eventually, however, the demand of maintaining such stringent controls across a spread of technologies becomes burdensome. One solution is to move to a looser, more forgiving approach. Another is to introduce specialization. As people expanded their horizons, it is likely that both solutions were used. By this perspective it seems inevitable that fluting would have been discarded in favor of strategies that were not nearly as restrictive. The question becomes "what was the final trigger that overthrew the long-standing fluting tradition?"

Once the lessons of science are revealed, it becomes clear that, with understanding, fracture is subject to considerable control. Since specific conditions lead directly to predictable results, the action of knapping need not appear haphazard. In the next chapter, we will explore how the correlation between cause and effect allows processes of making stone tools to be forensically reconstructed.

Chapter 5

Lithic Forensics

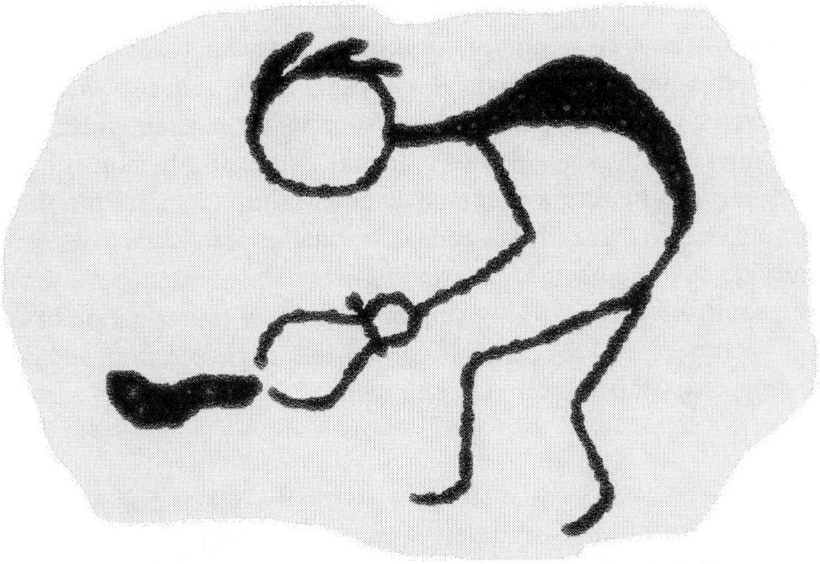

Forensic investigations require that past events be carefully reconstructed in a way that can stand the light of public scrutiny. Clues must be processed to paint a picture—in this case, of how ancient people went about their daily tasks and why they chose the technological solutions they did. An archaeologist gathers physical evidence very much like a criminologist. Both start with incomplete, often fragmentary data. They then must decide which pieces are relevant and what conclusions can be derived from them. If the job is done well, a meaningful and convincing story can be told about what happened in the past, and why.

Each fragment of stone at a site contributes information that must be analyzed to determine what people were doing there. From a modest sample, we should be able to figure out what tools were being used, and what strategies were applied to process raw stone. If we don't already know what the habits of a group were, the way tools were manipulated will be important in characterizing the cultural setting. It seems that the more we can learn about how craftsmen worked, the more specific questions we can ask and then answer. There is always a risk that not enough pieces of the puzzle are present to create a meaningful story. Ultimately, though, persistent investigation should form a clearer and clearer picture.

Most tool measurements reported from archaeological sites are standard data, collected in the hopes that it will be useful. An analyst can best design his study if he knows what information will answer the questions he is most interested in. Familiarity with a wide range of evidence from varied sources helps in anticipating the evidence, but is not always practical. For example, not everyone has the time to become an expert flintknapper. With that in mind, we will discuss several common characteristics of lithic materials that are useful for revealing what was most important to the people who left it. Many similar avenues of investigation have potential to expand the scope of knowledge available from scraps of stone, as will be seen in later chapters.

Certain knapping variables are particularly influenced by physics, such as angle of blow, platform offset, velocity of blow, preform support, and surface morphology. Since we were not present to record the conditions when an ancient tool was made, we must infer many of the variables from the outcome. Specific mechanics of flake removal leave diagnostic indicators. To the extent that these indicators are unique, they afford direct evidence to the habits and behavior of stoneworkers.

Physics shows that there is a vast range of ways to break stone. Humans, though, tend to find a familiar, comfortable, reliable set of conditions and then stay with them. Any knapper constantly compares the result of his action to his expectation. If the blow is too severe, he knows to ease up. If it is obviously misplaced, he can better direct the next one. This constant feedback and fine-tuning first leads an individual to improve his own techniques. Eventually, his descendants standardize the process into what we recognize as a shared tradition. On one hand, that crowds human activity into such similar modes that it is hard to separate one person's work from that of another in the same group. On the other, each tradition is likely to do things different from another because each has a unique history of flintknapping experience. The variability introduced by conscious decisions provides archaeologists with the means to identify subtle clues to behavior.

The way people perceive how nature operates accounts for at least some human behavior. By buttressing knapping experiments with computer modeling of fracture, we can see that repeating a successful action can lead to incremental improvements. Sometimes the behavior is based on a mistaken interpretation of nature, as when a knapper believes that swinging harder will make a flake go farther and discovers it does not reach as far as before. By astute observation, it is possible for an analyst to identify mistakes as well as successes.

When I first became interested in archaeology, I viewed flakes as just broken bits of rock left over from the making of some desired tool. But that is seldom the case. Each of those broken bits has a story of its own to tell. Sometimes a flake was made to serve a single function before being thrown away, but often its removal changed the shape of the parent rock to make a useful tool. That duality of purpose brings attention to the need to consider flakes independently from their remaining scars. Not only can the positive aspect of a flake convey valuable information, the negative scar helps to complete the data, and add new information about where the flake came from. There are obvious aspects of flakes that can be recognized and relatively easily used to interpret how they were made. In some cases, an expert's eye may be needed to decipher the deepest insights.

Flake shape

The way flakes are shaped is no accident. But what can we learn from that shape? In the previous chapter, cracks were modeled as a two-dimensional thin-section taken down the center of the flake. Since this is not what we see when we pick up a flake, another approach is needed to understand the third dimension of a flake, its outline. Isolated experimental flakes made on flat surfaces are a good place to start, since they have no surface topography to guide them. When off-margin impacts are used, the optimum flake shape is roughly a circle, tangent to the edge of the slab. The shape remains the same whether pressure or percussion is used to make the flake. The only substantial alteration of flake shape due to off-margin impact is when a poorly directed blow leaves the length short and causes a flattened oval scar. As soon as any deviant surface thickness is encountered, such as the bounding ridge from an earlier flake scar, or *arrise*, the flake advances most readily along the ridge, where thickness provides added stiffness. This means that flake shape is strongly guided by how surface topography of the preform adds or reduces stiffness around the perimeter of the flake. Otherwise, we would see only bulbous flakes of the same shape, distinguished only by size. Repeated flake shapes therefore indicate a deliberate repetition of the guiding surface geometry. Controlled experiments by Peter Kelterborn (2002) establish conclusively that while flakes retain a consistent shape for a given surface morphology, their size is proportional to the longitudinal force of detachment.

Any part of the surface that stands above the rest causes stiffness that steers and extends flake travel. Flakes are seldom made on perfectly flat surfaces, and ridges of only tenths of a millimeter are sufficient to contain their expansion and allow extensive travel. The careful contouring of a preform encourages a flute to travel in a predictable path along the deliberate crest. On the other hand, when knappers want to make a flaring flake scar, they need to avoid distinct ridges. If a collection of flakes each tends to expand uniformly, they likely represent a stage of work where uniform thickness was a priority and ridges were subdued. Any time a shape is repeated consistently, it means flakes were deliberately made under the same conditions. If we can figure out why those conditions were chosen, we have effectively looked into the mind of the maker.

When Levallois craftsmen domed their cores, it provided nearly equal stiffness for every part of the flake. Width-to-length ratios of spent Levallois cores show that they correspond to the proportion of an ideal flake from a flat surface. Other core shapes would have been easy to make, but none could be so reliable. The strategy also consistently produces the widest sharp-edged flakes.

Flake shape

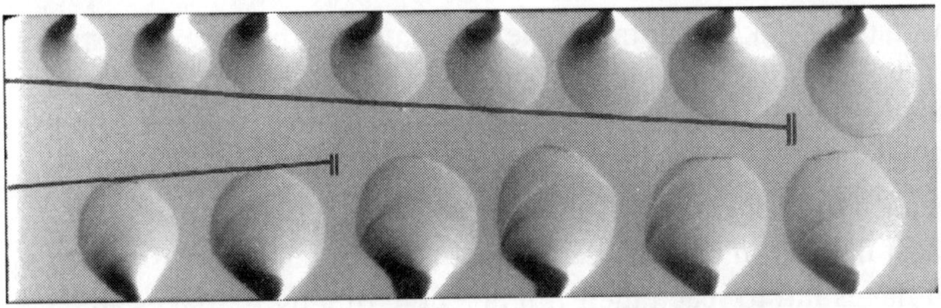

Because flakes made on plate glass do not have ridges to guide them, they are nearly round. The size difference in this case is due to changing the longitudinal force. For a given surface, flake shape remains the same whether made by pressure or percussion.

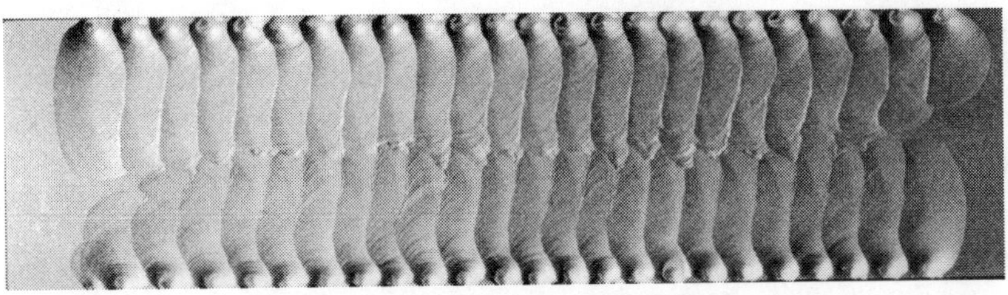

A flake reaches farthest when it has the stiffness of a ridge to guide it. Subsequent flakes have similar shapes as long as they follow equivalent ridges.

These flakes were made by Peter Kelterborn under precisely controlled conditions. By permission, from Measurable Flintknapping (2002)

Flake extent

The distance a flake travels depends on how the core is supported. The flake's intended purpose can be inferred from how far the flake reaches. Because flakes are so useful for cutting, the practice of making flakes that travel the full length of the core dates from very early times. The longer a flake is, the more cutting edge is available. It still took thousands of years to progress from thick flake cores to thin tool preforms. The highly specific conditions needed to make an edge-to-edge flake are so hard to meet that most flakes can be considered fractional. In fact, the vast majority of flakes fail to reach their full potential of length. Flakes are usually made intentionally short, although they may be fractional by accident. Without the corresponding core and scar, it can be difficult to recognize to what degree a flake traveled its potential length. Step-flakes, hinge flakes, and overshots are easier to recognize because they fail to form a sharp, acute edge over their full potential length.

Step terminations, where flakes are broken off squarely, are almost always caused by error. It may be the result of a weak swing, use of too light a hammer, or a yielding support. When a flake tries to progress beneath a lump on the preform surface, it may run out of energy and cause an especially severe step flake. The extra thickness of a flake also generates a bending force that makes it prone to snap off rather than progress. Hinge flakes, which terminate in a tight curve, reveal another kind of error caused by using too stiff a hammer. Overshots almost always result from bending the preform by mis-aligned support while trying to make an edge-to-edge flake.

Routine creation of edge-to-edge flakes reveals that process controls are in place to assure success. When the support provided is not sufficient to send flakes edge-to-edge, the resulting flakes are shorter and feathered. Over time, a knapper's muscles become trained to allow predictable work. Those ingrained habits also help him avoid serious mistakes, and consistently produce the same kind of flakes. Feathered fluting scars should seldom occur on fully-fluted projectile point styles because the support necessary

Flake extent

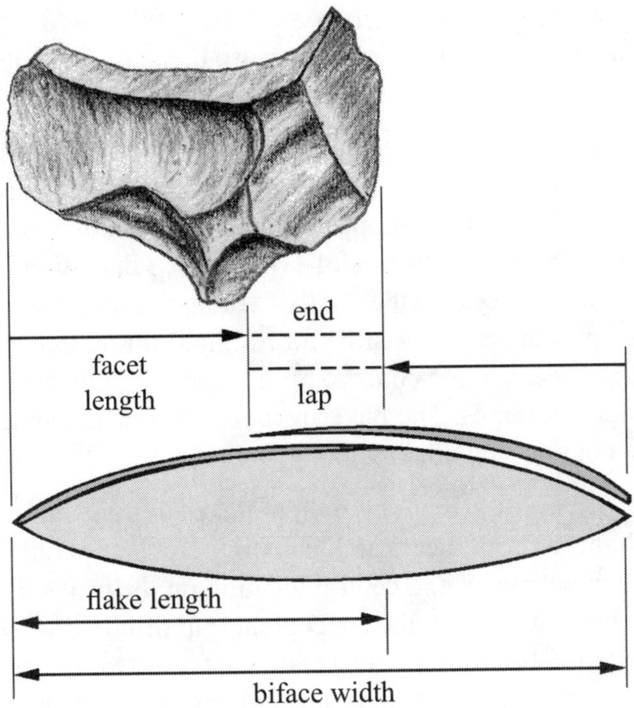

Flake length = start-to-finish length of a flake.

End lap = maximum extent of scar from opposing edge.

Facet length = (flake length) - (end lap)

Biface width = (2 x flake length) - (end lap)

% End lap = (end lap) / (facet length)

% Flake extent = (flake length) / (biface width)

By definition, full-length bifacial flakes have 100% flake extent and can have up to 100% end lap. In defining processes, we need to identify meaningful trends among quantities of flakes. Analyzing flakes from specific stages of reduction helps quantify the rate of biface narrowing.

for creating flutes prevents the fracture from exiting the face. Conversely, early Clovis points were seldom supported well enough for flakes to reach the tip because the goal was simply to make a wedge-shaped base that could be easily hafted.

Face flattening is a strategy that involves minimizing the relief of a face by taking either lateral or longitudinal flakes. Earliest use of this principle was the European Levallois technique, where the flake, not the core, was the goal. Levallois face flattening vaguely resembles fluting because it actually exploits a set of conditions on the doorstep of those responsible for fluting. Fluting in Clovis and Folsom times carried the concept of face flattening to its logical and physical extremes. The flakes responsible for flattening are generally edge-to-edge flakes.

Fluting involves using longitudinal flakes to thin a biface, taper its base, or both. Scars are distinctive by their parallel edges, blade-like appearance, and central location on the projectile face. It is easy to fall into the conceptual trap of thinking that all flutes needed to be full-length flakes. In actuality, only a few rare projectiles were made with flakes reaching the maximum extent on both faces. Even following the process does not guarantee a perfect outcome because any small imperfection in the stone can shorten the flute.

As stated earlier, a flake can be used as a tool of itself, and/or as part of the shaping of another implement. If both ends are to be satisfied simultaneously, the knapper always prioritizes one or the other goal. Flake by flake, we may not understand what was intended, but when many flakes and flake scars are considered, the trends can become evident. Generally, edge-to-edge flakes destined for tool use are thicker than fractional flakes that merely contour the surface of a biface because added preform support forces the flake to cut deep under the surface.

Flake flatness

A flake is flat or curved, depending in large part on how the core is supported. Since people do things the way they are taught, the arc of a flake provides evidence of their knapping habits and may help us to differentiate one group's work from another.

Depending on what a knapper wants to do, fracture surfaces may range from plane/flat to arc/curve. In general, flat fractures show that the natural tendency of the stone to bend has been overcome. Restraint of a core, whether through added mass or grip, leads to flattened flake trajectories. Increasing flake thickness also adds stiffness and lowers curvature. Supporting the preform against an anvil, necessarily moves the center of mass toward the core tip, and flattens flake curvature most effectively.

To encourage curvature as a flake progresses, the support needs to allow movement of the preform. When the core is allowed to rotate, the removal force arcs about the center of mass and develops curved blades. When the preform is allowed flexibility, the use of a yielding antler baton is most appropriate. Clovis overshot flaking on bifaces is most readily replicated by pinching the preform at the far edge to make the preform act flexibly, and exaggerate the arc of flake travel. Greater bending influence leads to increased curvature as the fracture proceeds. While flake curvature does not offer distinctive separation of flaking methods by itself, combined with other diagnostic features, it allows an analyst to judge the means of support with reasonable confidence.

The faces of a Folsom point are usually so flat that the point can lie on a table without rocking. Spiraling flutes would cause one corner of the point to raise off the surface plane. Flutes that arc from end-to-end would allow one end or the other to rock away from the table surface. Very flat flakes common to Folsom bifaces are encouraged by supporting the whole perimeter to restrain any rotation of the preform while the blow is delivered. Only percussion seems to consistently produce projectile faces sufficiently plane to match what is found archaeologically among fluting traditions. Fluting traditions were especially adept at varying support to exploit the spectrum of curvature. Most other traditions stayed with a single system of support.

Bulb of force

Knapping skill depends on blows being directed properly in relation the how a core is supported. The bulb of force helps reveal a knapper's expertise in setting the angle of blow. Australian researchers, Brian Cotterell and J. Kamminga (1987), used engineering beam theory to derive equations for the angle of force required to make a crack grow parallel to the core face. The crack must start in tension, with a 90-degree angle of force, but very quickly requires an angle of force that nearly parallels the core face. They point out that the stiffness of the developing flake acts to keep the force angle on track when the angle of blow is between 0 and 60 degrees. The angle of blow (at first contact) does not change and always differs from the angle of force, which varies throughout the flake duration.

The graph on the following page shows that the angle of force must decline as the flake progresses and suggests that starting with low angles of blow makes it harder for the flake to adjust to its final, near axial, loading condition. No matter what the starting angle of force or platform offset, the adjustment takes place in the same distance and manifests as the bulb of force. Computer modeling indicates that axial impacts cause more severe bulbs of force than angled blows. In effect, the graph describes a governor effect, where the flake responds to an applied force and attempts to keep on its appointed path by changing its thickness. Therefore, when the starting angle of force is too low, the bulb of force increases thickness until the crack is stable. A perfect blow would leave virtually no discernable bulb of force.

Fast loading rates associated with increasing hammer stiffness exaggerate the bulb of force. Flaking from a beveled edge effectively minimizes the bulb of force while maintaining directional stability, but a high angle of blow helps as well. After a very short distance of travel, the fracture path stabilizes and is very resistant to external steering.

Fluting by direct percussion leaves a more subtle bulb of force than punch or pressure. Baton orientation helps greatly by increasing the effective angle of blow. Bulbs of force are seldom

Angle of force required for crack to parallel core face.

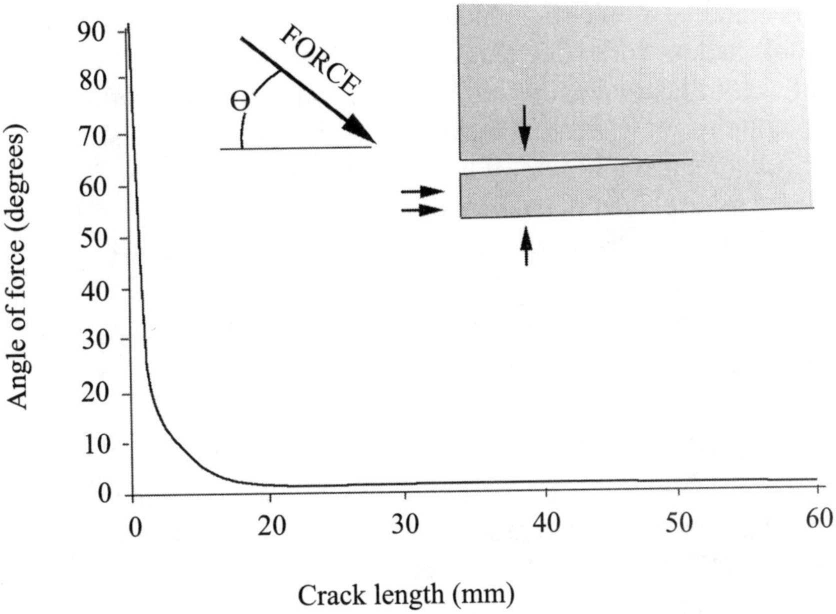

A bulb of force is created when the initial angle of blow does not match the angle needed to keep the crack straight. Low angles of blow cause the greatest bulbs of force.

Flake cross section showing the bulb of force.

distinct in Folsom technology and are key to achieving the lower limits of Folsom thickness.

Flake thickness

Flake thickness tells something about the minimum size of cores people had to work with. But what limits are placed on thickness of flakes? Flake thickness has so much to do with core surface topography and platform placement that it can serve as a signature of technological tradition. While the edges of the original flake get altered by use, the thickness at the center remains unchanged, and can attest to how large a core it came from. Whether a flake serves as a tool blank or reduces the core surface, it is still subject to the same rules of fracture physics. Most evidence remaining at sites is reduction debris because many of the large flakes were taken away to be used as tools.

The diagram on the following page relates the thickness of a core or preform to the greatest thickness of flake that can be struck from that size core. Two ratios of flake thickness (t) over core thickness (T) are drawn for reference. Making a flake half the core thickness is most difficult and unlikely because it means splitting the core. A more practical limit is illustrated by Folsom flutes that seldom exceed two fifths the thickness of the preform. In principle, this diagram can represent flakes running end-to-end on a core, or edge-to-edge on a biface. As core size decreases, an increasingly accurate blow is required to control the fraction of material removed.

To achieve maximum reduction of a core by edge-to-edge flakes requires increasing precision of successive blows. By the time fluting is attempted, there is hardly enough core thickness to allow the task. Because it is so difficult to remove a flake more than 0.4 the core thickness, an edge-to-edge flake found at an archaeological site had to come from a core or preform at least 2.5 times as thick as the flake. As scrapers are often made on edge-to-edge flakes, they give excellent clues about what size cores were in use. It doesn't matter if the flake has been re-worked on all edges—thickness is the key. This principle of how flakes are related to cores will be revisited more than once as we consider the lessons of archaeology.

The relationship between flake thickness and length provides a valuable indicator of the original state of broken and reworked flakes. Experimental work by Pelcin (1996), which is supported by Baker's computer model, shows that the mass of a flake is fixed by the conditions necessary for its formation. As a consequence,

Geometric relationship of flake to core thickness

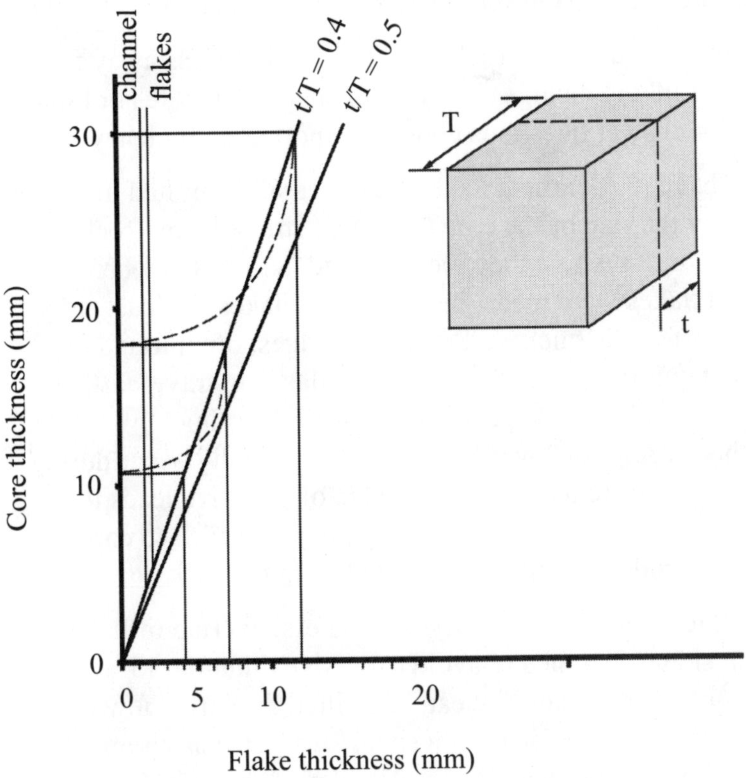

The Folsom flake to core ratio of 0.4 is used as a practical limit, although a ratio of 0.5 represents splitting a core in half. Each flake removal leaves a smaller core and lowers the potential thickness of the next flake. Flake thickness, along with other clues, can signal knapping tool preference and intended use of the flake.

measurements of bulbar thickness may be used to extrapolate original flake mass and length with fair reliability.

It is difficult to assure that all possible flake conditions are represented by comparing modern knapping products with archaeological data. The computer model of fluting, however, allows the mapping and comparison of every potential flake path. Measurements of archaeologically recovered channel flakes average 2.2 mm, with a standard deviation of 0.36 mm. The agreement with the computed theoretical median thickness on page 98 is remarkable.

Defining the production sequence

Flakes can show us how a rock is reduced as it travels away from the quarry. Linking changes in flakes to changes in cores or bifaces allows us to answer questions about transport and efficiency.

As we have just learned, flakes are necessarily limited in thickness by the size of the core they were struck from. Rocks are always largest where they are collected, and that is where the biggest flakes were made, because large flakes are least risky when the preform is thickest. Prominent ridges, often found on unmodified quarry rocks, also encourage flakes to travel farther. Edge-to-edge flakes are especially useful for learning how wide or long the starting rock was because they generally occur during the early stages of reduction. Knowledge of core size leads to estimates of what size pack loads early people needed to carry and what usable products could be extracted from the rock.

As a biface approaches its target thickness, overlap of flakes from each side is lessened to avoid excessive thinning. Ideally, biface remnants, or flakes that can be refitted directly, show the typical width of bifaces made at a site. However, measurements of debitage allow reconstruction of a typical preform cross-section from an estimate of the relative extent of normal flakes.

Dimensional data about flakes for a site may allow several avenues of analysis. Not only can we learn how a biface changes shape, we can define the range of variation that is allowed for each stage. With a good understanding of the overall process, specific phases of work can be identified at sites with limited activity. Once typical flake extent has been defined for each stage of work, the decision-making and knapping practice unique to each culture can be distinguished. By contrasting inferred large flakes with those that remain, the initial flake tool requirements become evident and allow us to estimate the degree of modification represented in abandoned tool sets.

As the histogram on the following page shows, most Folsom reduction flakes probably came from bifaces not much thicker than 20-mm. But tool flakes are typically too thick to have been made from a core less than 15-mm thick. We will revisit this issue in

Frequency of flake thickness

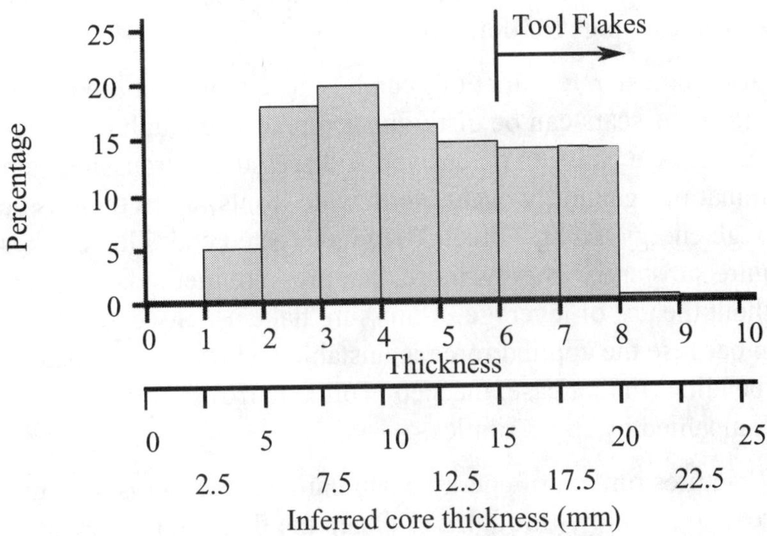

Frequency of thickness of flakes from the Antelope Spring site in Colorado, shows that 80% of total flakes are under 8-mm thick, and could be made from bifaces under 20-mm thick. Nearly 95% of tools at the Stewart's Cattle Guard site are thicker than 6-mm, and should have come from bifaces thicker than 15-mm.

much more detail in chapter 7 when we investigate the life cycle of a rock.

Terminations

The way flakes terminate can tell us a lot about how knappers wielded their tools. Workers familiar with the mechanical conditions responsible for terminating a flake can use the knowledge in a feedback loop to adjust their blows and platforms to optimize their success. Step flakes are often caused by a combination of low blow angle and low platform offset, leaving insufficient energy for the flake to complete. Follow-through of the knapping tool can then break the flake off short. Feathered terminations are associated with relatively high blow angles and thick platform offsets, but may also indicate the absence of anvil resistance. When blows are directed too parallel to the face, it can cause overshots, especially when the preform has strong support. In an extended work session, there is ample opportunity to adjust blows and support for optimal effect.

Although step terminations can be caused by most modes of fracture, the scars can be distinguished readily enough to recognize the difference between percussion and pressure. Percussion step terminations generally occur along wide fronts, when there is not enough energy to carry the flake to its targeted end. Thick flakes require substantial energy, more than pressure methods can achieve without the use of leverage. A pressure flake usually ends in a step because the applied force is unstable and breaks the flake off by bending. In that case, the step is often narrow, shallow, and accompanied by slight ripples.

Fractures run out of energy when hammer contact is lost due to rebound. Follow-through can't keep the flake on its previous course and snaps the end as a step. Understanding this, we can recognize step breaks in thin terminations—those under a millimeter—as representing pressure flakes because pressure uses the lowest potential energy. This provides a useful way to separate pressure retouch scars from earlier stage percussion work.

A hinge flake is caused when the loading time is too short, and the angle of force sends the fracture curving to the free surface of the core. To qualify as a hinge termination, the flake executes an abrupt but tight radius turn, to exit the face at 90-degrees or more. Rippling at termination could be mistaken for hinging except for the angle of exit and the fact that ripples are progressively smaller, while hinges are exaggerated. Since hinges occur when loading time is either 0.15 or 0.85 of the core's natural vibration frequency, the length of the flake becomes an important clue that the hammer rebounded too fast. With fast rebound, the flake runs out of lengthwise push and has no option but a quick trip to the face. So, a flake that hinges immediately has been hit with an especially stiff hammer. Flakes that hinge well away from the point of impact have fracture time just short of the upper threshold. No hinges should be possible with the slow loading of pressure flaking, and organic punches or hammers should minimize hinging.

Platform remnant

The accuracy of a flintworker's blow can be inferred from subtle details of the portion of the striking platform that remains with the detached flake. Close examination of flake initiations shows that Folsom flakes typically retain only a slight remnant of the platform surface that was struck. This characteristic can be most easily and consistently reproduced by controlling where the direct percussion blow lands, to avoid overstrikes. When the blow is not precisely directed, we should expect to see platform remnants of many sizes. Freehand percussion without anvil support tends to allow varying degrees of contact with the platform surface, causing varying amounts of surface to be carried away with the flake. Punches can remove more of the platform if the punch tip is positioned away from the edge, but can duplicate direct percussion when the edge of the punch just engages the platform.

Because Folsom flakes retain so little of the platform surface, it shows that the workers dismantled stone more carefully than most later people. A less conservative practice would exhaust the rock too quickly. In areas where stone was abundant, people tended to use less demanding techniques since a replacement rock was readily at hand if mistakes were made. Debitage found where stone was plentiful show that impacts far from the edge were common, and errors of impact placement were also more frequent.

Platforms are important in determining how strong a blow was required to initiate the crack. The success of any knapping tool depends on building the platform so that it lets a crack start at an appropriate level of force. Too much force can destroy a platform, and too little will not form the right flake. A platform suitable for an antler pressure tool to set against would be destroyed by a copper tip. The adherents of any tradition can be expected to use a standard tool set in a relatively uniform manner. Therefore, when speculating about tool requirements, platform characteristics should be examined very carefully. Keep in mind that a sturdy-looking platform on brittle stone like obsidian might require less energy to release than would a delicate platform on a tougher material such as chert or flint.

Platform offset

Strategy and technique are intangible concepts that can be inferred from the way impact sites are chosen and constructed. Folsom bifacing flakes can be sorted by their relative thickness, which indicates how far impact was from the face plane. Thin flakes have impact sites close to the flake surface. Thick flakes have platforms offset considerably. Either type of flake can be removed by using the same support and angle of blow.

Thin flakes often have carefully isolated and ground platforms that are near the unmodified biface surface. Because the offset needs to be precise, the platforms are usually set up by pressure. These flakes are not destined to be used as tools, but to sculpt the contour of the biface. Thin flakes that look like potato chips probably represent refinement of bifaces that are nearly ready for pressure refinement.

Platforms of thick flakes are often deliberately offset from the original biface surface by small step-flakes caused by crushing or battering the edge. The impact sites are usually near the thickest portion of the biface in order to both to make a thicker flake and to quickly thin the preform. Thick flakes are useful to create bifacial blanks for later alteration, or unifacial flakes to use as camp tools. Another purpose may be economy of effort since deeply-penetrating flakes get the job done with fewer strokes. Since this category of flake usually represents the earliest phase of reduction, it often exhibits a large amount of cortex removal.

Sources of ripples

Since an ideal flake is free of ripples their presence on a flake can tell where a knapper strayed. A correctly struck flake should be nearly free of ripples because the knapping force remains properly directed for the length of the flake. A number of factors can send the crack off course. Nature uses material stiffness as a sort of governor that tries to put the fracture path where it belongs. Once the flake starts going off course, like a race driver catching a wheel in a rut and then overcompensating to get back on line, a series of undulations form that are longest at the outset and spaced closer as they reach the end, making the troughs seem deeper.

Ripples are particularly common where the flake narrows rapidly. A sudden reduction in flake width changes the stiffness of the flake, and the crack has to undulate to seek stability. Stronger stone or deeper flakes lower the effect of bending and reduce the incidence of ripples. Flakes that travel fast also minimize rippling. Relatively slow flakes made by pressure allow the most ripples, especially if the preform surface is not absolutely uniform.

Changes in the thickness of a flake are usually responsible for substantial alterations in the direction of force that make the largest ripples. Variations in surface topography are often mirrored in the fracture surface because stiffness of the detaching flake guides the depth of the flake path. Fluted points are known for having subdued rippling in the flute channel. Ripples occur because uniform retouch flakes found on Folsom preforms cause regular intervals of surface bumps and hollows that are mirrored on the bottom of channel flakes. Folsom workers tended to use wide flake spacing while preparing for first flutes, making broadly spaced undulations likely on these flutes. Relatively close pressure spacing for second Folsom flutes may have helped subdue undulations when the preform is especially flexible.

Ripples can seldom be completely avoided, but Folsom points usually have especially gentle undulations, rather like ocean swells. Choppy ripples indicate erratic bending forces, and are unusual in fluting technology because the process is so well controlled. Direct percussion can cause rippled surfaces if it is used improperly, but it can subdue undulations better than most other means of fluting because of the high speed of fracture

Not all ripples reflect major corrections to the flake path. Anything that interrupts the flake's progress can cause a pulse-like interference. Imbedded flaws in the rock can create low amplitude ripples that look like tiny lines, but many ripples result from damage to the knapping tool or the platform. Crisply defined ripples that do not have a recognizable interval are caused by interruptions too brief to change the overall flake path. Small, chattered ripples, left when the platform is damaged, can sometimes indicate when hard hammers were used.

Fracture Velocity

Fracture obviously happens quickly, but just how quickly? Measuring fracture speed after the fact is possible under special conditions and provides clues to how the blow was delivered. Since the speed of fracture is proportional to the rate of force application, knowing the velocity of a fracture, could tell about the manner in which force was applied. But how can the speed of fracture be measured thousands of years later? Direct measurements can be made by adhering electrical circuits to the stone surface and recording when fracture disrupts the contacts. High-speed photography can also be used to observe a fracture in progress. Don Crabtree (1968:470-476), an early pioneer in flintknapping studies, used high speed photography to record the removal of obsidian blades by the pressure-crutch method at 125 meters/second. That velocity of crack propagation reflects a relatively static loading mode. But how do we determine speed of fracture that occurred in ancient times?

Measuring the speed of a flute

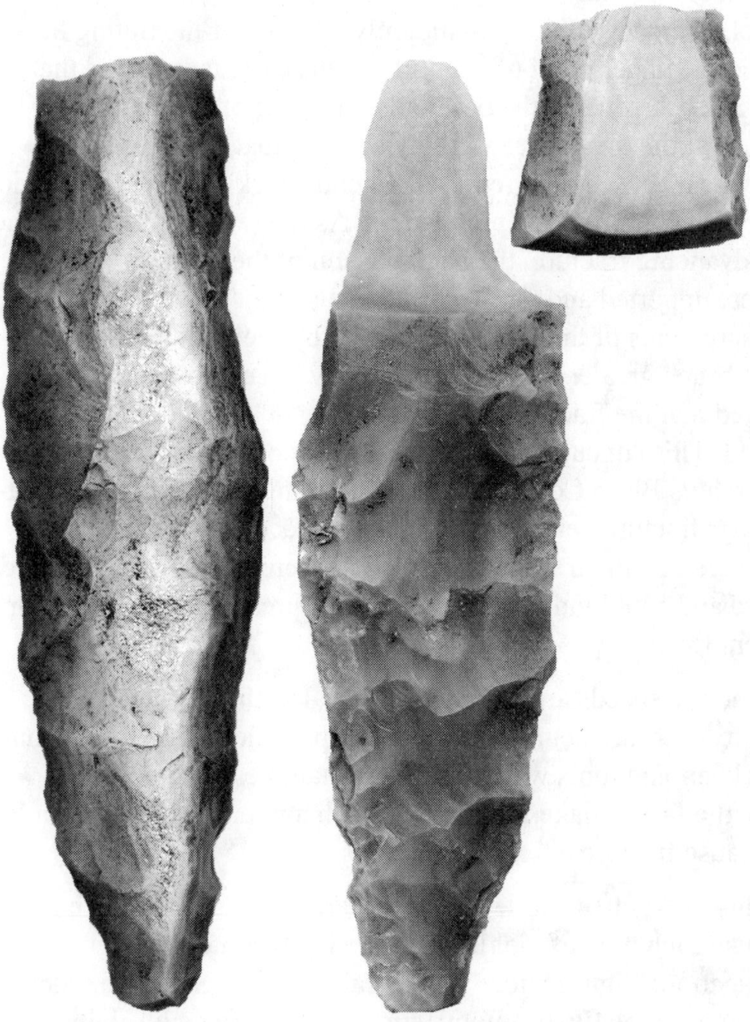

This break occured when the compression wave from the blow traveled to the tip and back, and finally severed the base by reverse end-shock. Since the speed of compression waves in flint is known, the speed of fracture can be calculated to be about 760 meters per second.

My own experiments with percussion fluting using an antler baton, with tip support against an antler anvil, a failure occurred that should not have been possible. According to the computer model, overshot failures should only occur after the fluting flake passed the upper point of support. In this case, it appeared that an overshot had severed the base only a quarter of the distance to the support at the tip. Closer scrutiny showed that the "overshot" was not typical and, in fact, represented end shock from a compression wave reflected from the preform tip. As the reflected wave reached the advancing fracture, the back portion of the preform was no longer supported and tension stress caused the section to fall out. Measurements of the distance traveled by the fracture relative to the much greater distance traveled by the compression wave showed that the fracture was progressing at about 760 meters/second. This concurs with direct observations by Australian researchers Brian Cotterell and J. Kamminga (1987:681), of hard-hammer fracture velocity at 800 meters/second in chalcedony. Evidence of similar failures may be expected in the archaeological record and could indicate the conditions present during fluting in ancient times.

Fracture speeds are linked to material stiffness, therefore velocity measurements can be used to provide information about what knapping tools were used. Hard hammers will generally cause the fastest flakes, while organic hammers or pressure tools will cause fracture at lower velocities.

This means that subtle signatures are apt to exist in flake scars because each type of tool leads to distinctive rates of fracture propagation. High fracture speeds can cut through material defects to produce a relatively smooth surface compared with slow fractures that leave a rougher surface.

Breakage patterns during fluting

Mistakes are inevitable, but when a mistake is made, it preserves a phase of work unaltered. The nature of the error can sometimes reveal a knapper's technique or strategy.

The method of applying load during fluting may cause characteristic breakage patterns. Since archaeological examples of overshot occur primarily near the tip, the support is assumed to be near the tip, or at least distributed evenly enough for overshots to occur some distance from the impact site. The relatively common incidence of tips being snapped after the final flute was taken may be interpreted in more than one way. Follow-through of the fluting stroke can bend the preform in two with end-to-end load after the flute has detached. Bending during retouch can snap the preform in two when the center is left unsupported, like a weak bridge. However, the majority of snapped tips lack signs of unfinished retouch. It has been speculated that the tips were deliberately snapped to avoid having to thin them, but this seems an unlikely waste of length and thinning is too easy to *not* do it.

Split preforms and burinated ears can occur during pressure fluting, but are much more likely to happen during percussion. Static loads make spiraled flutes more likely, so the fact that channel flakes stay flat is another good indication that dynamic loading was used. Because of the slight clearance between the fluting nipple and the ears, it seems reasonable to expect ears to be crushed or splintered during percussion fluting. A low incidence of ear damage may be explained by a high degree of control over the blow. Alternatively, many ear contacts may have resulted in splits, burinations, or damage that was simply repaired by retouch. Trials show that by placing thumb and forefinger of the supporting hand at either ear, the hammer can be gently deflected without smashing the preform ear—or the fingers either, for that matter.

Buckling is represented archaeologically by right-angle bend breaks. While any method of fluting can produce buckling, we would expect the highest incidence if pressure were the predominate mode, because holding the preform in place with a vise causes bending. Most buckled failures appear to have occurred while pressure flaking the preform in preparation for the second flute. Supporting the preform at the ends and not the middle can cause it to snap in two. If the manner of fluting were the primary cause of buckling, failures during the second flute attempt should be more common than during preparation.

Breakage patterns

Various kinds of production damage from Folsom sites include:

A. Snapped tip, with flute scar remnant (Antelope Spring).

B. Overshot tip (Bobtail Wolf).

C. Overhot base (Bobtail Wolf).

D. Split preform [both faces] (Antelope Spring).

E. Burinated ear on replica made by Bob Patten.

F. Failed 2nd flute associated with simultaneous ear shatter on right (Bobtail Wolf).

G. Snapped base (Antelope Spring).

Differences between pressure and percussion

Completely different actions are required to shape stone by pressure than by percussion. While difficult to decipher, there are clues that may tell the difference.

Distinguishing pressure activity from percussion is important in defining stages of work. In general, percussion is used for quick and easy coarse shaping, and pressure is reserved for refining. When were modes changed, and why? I know from personal knapping experience that a pressure flake can be made to look very much like one made by percussion. I also know that a knapper switches from percussion to pressure when it makes work easier rather than because it is attractive. When knapping actions can be relegated to specific stages of making an artifact, it should prompt us to look for the reasons for the shift.

Mathematical modeling describes pressure and percussion flaking equally well. The difference stems primarily from how each technique stores compressive energy. Pressure allows the knapping tool to be fully compressed when the crack starts, but requires that the core be firmly supported by a vice or simply by hand. The more flexure present in the presser, the farther the flake can progress. A percussor can never be completely compressed before the crack begins. Still, the ability of the hammer to store energy is critical to how far it can send a crack.

There may be no single clue that conclusively tells whether percussion or pressure was used, but a combination of indicators can be persuasive. Unless massive leverage is used, pressure flakes must be considerably smaller than those normally produced by percussion. Hand-held support for pressure also allows the flake path to waver much more than it would for percussion flakes, partly because the fracture takes longer to complete. Pressure retouch flakes are not only small, they have a standard size—virtually impossible to achieve by percussion. Also, pressure easily accommodates patterned flaking that cannot be accomplished with percussion. As mentioned earlier, hinge terminations are rarely, if ever, associated with pressure because fracture is caused only by the static rebound of the core.

Putting the processes in order

Exquisitely formed artifacts make it clear that knappers could do what they set out to do, on demand. Backtracking what was done reveals the intent behind the action, and whether the worker was guided by a master plan.

Particularly when dealing with fluting technologies, we can safely assume that each flake was made according to a deliberate plan. Cores were not smashed apart indiscriminately so that the debris could be scavenged for pieces that just happened to suit a purpose. A knapper made each flake in anticipation of its use as a tool, or as a definite improvement in the contour of the core. Only when we understand what was intended can we deduce whether the process was wasteful or capable of extracting the maximum utility from a quarry blank.

If the strategic role of each flake can be identified, its place in the reduction process is made clear. Since more than one purpose can be met simultaneously, assigning a purpose to every flake is problematic. We will content ourselves, for now, with identifying the highest priority for each of three broad phases of activity.

Quarrying, roughing-out, and refining are reasonably easy to differentiate by size and flake patterning. While some overlap may be expected among these activities, they typically are not conducted at the same sitting for a single starting rock, or even at the same location. In analyzing debitage, we can use our knowledge of how each phase of dismantling a rock leads to characteristic morphologies.

During quarrying, the earliest flakes are generally designed to remove defects such as cracks, inclusions, and cortex. This previewing determines whether more effort is worthwhile, or if the rock will be abandoned. Large rocks need to be broken down into relatively sturdy tool blanks between 17 and 55 millimeters thick. Blanks are typically long, broad sections made by stone-hammer percussion. In preparation for transport, sharp edges, thin protrusions, and thick humps are likely to be removed when they do not add to the stone's potential for yielding tools.

It can require very few blows to transform the unstructured shape of a blank to a biface. Initial rough-out flakes seldom reveal much about the initial shape of a quarry blank, and cortex is common. Since thick preforms most readily yield thick flakes, scrapers are likely to retain cortex from the quarry blank. Scrapers are also least apt to demonstrate whether they originated from a biface or not, because they are heavily modified. Flakes need to be greater than 6 millimeters thick to make serviceable camp tools, so the core generally needs to be thicker than 15 mm.

On the way to becoming a biface, a blank often presents a steep or abrupt edge that can be very difficult for the flintknapper to deal with. One solution removes the corner with blade-like flakes. Another uses shorter flakes aimed obliquely from an adjoining edge. This technique of flaking alternately from face-to-face, while creating an edge, is referred to as sewing. Finally, a flake can sometimes be sent from the opposing edge to remove the blunt edge by overshot.

Each of these strategies for establishing a biface edge leaves characteristic flakes. Corner blades are very different from blades taken in sequence from a core because their dorsal faces are merely faceted breaks. Forming an edge by sewing causes highly asymmetrical flakes that often retain some of the offending abrupt edge on the high side of the platform. Overshot flakes meant to solve an edge problem very often preserve that steep edge remnant as well. Fortunately for us, these kinds of atypical flakes were not highly desirable for tool use and usually remain on site, unmodified. Relative frequencies of such distinctive flake characteristics can tell an analyst whether the starting morphology was typically rounded or squared.

Once a biface has been fully edged and given shape, it has less value as a source of heavy tool flakes. At this phase, thinning and contouring become the highest priority. Both kinds of flakes tend to be symmetrical because the preform surface has already been regularized. Thinning flakes trend toward flatter curvature and normally reach well past the preform center, as indicated by scars on the dorsal face. Contouring flakes may not reach the preform midline and usually are narrower and more delicate than thinning flakes.

Pressure flakes are usually produced during finishing and rejuvenation. The debitage is usually so small that it is difficult to analyze. Pressure flakes can exhibit all the characteristics of percussion flakes, because the same laws of physics guide their formation. Their small size makes them especially vulnerable to being blown or washed away from the site. Some of the flakes are produced en-masse by shearing edges with a single rasping swipe, and may show no platform preparation. Pressure flakes can occur from any part of the production process, but are most numerous while refining projectiles and sharpening camp tools. If a site has sufficient preservation and definition, it may be possible to associate small flakes with activities like preform refinement or tool maintenance.

Refitting

Refitting allows an analyst retrace the production timeline to see what organization sequence went into building an artifact. Fundamental insights into the knapper's goals, strategies, and techniques are most clearly revealed when flakes can be refitted onto an artifact. Through refitting, we can understand the dimensions of early stage preforms and the movement of stone through a site as sequential activities contributed to the dismantling and distribution.

While refitting is a powerful tool for understanding subtle nuances of lithic reduction, it involves some problems as well. The original workman did not anticipate that a future archaeologist might want all the debitage left intact and in pristine condition. Pieces of the puzzle are almost always missing. Some are lost, some are used for tools, and some are left at other sites. Certain knapping techniques leave no recognizable trace of a flake. Even that, however, gives us knowledge about the workman.

The more one works at refitting, the more it becomes obvious that the "negative" missing parts are at least as important as the "positive" fits. Refitting often provides critical links between different work areas of a site. Finding a piece apart from the primary work area can indicate that physically separated activities took place as part of a single event. As an example, matching the tip of a broken scraper to the remainder found at a separate hearth area shows that the two locations were contemporaneous.

Refitting studies provide an interesting contrast between Solutrian and Clovis technology. Only isolated flake sequences can be refitted from Clovis sites because most large flakes were removed for use as camp tools. At the Les Maitreaux quarry site in France, enough flakes remain from reducing large Solutrian bifaces that the entire outer shell can often be reconstructed.

The best way to learn about refitting is to create a modern replica, then fit the flakes back on it. Even with all the pieces present, it is no trivial task. If you wish to attempt it yourself, the following list shows some things that can make the job easier:

- Start by fitting broken flakes back together to make options more recognizable and reduce the number of flakes.
- Work from the inside out, especially when fitting to the artifact.
- Landmarks of color, inclusions, and contour allow fitting complete flakes. Fit is usually confirmed by feel of the flake locking into place.
- Gaps left by missing pieces should remain unglued until there is no doubt that the fragments are unrecoverable.
- End-to-end flake overlaps help define the span of the original artifact. Once width has been spanned, additional fits are easier to locate.

Archaeological refitting projects seldom allow neat, unambiguous reconstruction, but they do point to patterns that can be substantiated by other clues. Matching rebuilt replications to the archaeological sequence helps to demonstrate that the process is properly represented. (see page 173)

Knapping tools

Only rarely do we find tools used by aboriginal knappers. Even without them, there are ways to determine what they were and how they were manipulated. By definition, a flintknapping tradition trains its practitioners to use a standard tool set in a prescribed manner. There may be no single characteristic that testifies clearly to what kinds of tools were used, but the more indicators there are, the better the determination.

Consistently small platform remnants point to a precise impact location that can result from mechanical control of a baton or a rocker punch. Folsom flakes often have less than a millimeter of platform surface removed with the flake. This implies the use of antler knapping tools because harder tools would have damaged such delicate platforms.

Ideally, platforms are built to be just as strong as they need to be for full flake development, but not so strong as to resist the crack initiation. Stone hammers require stout platforms to avoid destruction of the impact site before the crack can initiate. Antler punches or pressers can be successful with weaker platforms, provided relatively short flakes are acceptable. Experiments show that fluting traditions used platforms appropriate for striking by antler batons.

The fracture initiation, viewed in line with the blow, reveals how the force of impact was distributed. Soft hammers leave an arced (*arcuate*) detachment signature, while hard hammers cause a wing-shaped (*alar*) detachment because of the small Hertzian component associated with hard impact. The low number of alar detachments found with fluting traditions makes the use of hard hammers for biface refinement appear unlikely. The fact that the smallest platforms are usually found intact supports the use of soft hammers rather than hard impactors.

Interpreting data

The human mind is a powerful instrument for processing data, but statistical resolution of meticulously collected data helps to provide objectivity. As surely as people perform with intent, they also fail to hit the same mark each time, and that can leave clues that sometimes point in misleading ways. The fact that processes and stages exist means that workers carried a mental template of what sort of flake was most ideal for a task. Meticulous evaluation of archaeological measurement can show that a purpose was being satisfied by demonstrating how well the data clusters to a distinct pattern.

If only one goal existed for the worker, plotted frequencies of a pertinent variable show a bell-shaped distribution curve. Multiple goals, however should appear as separate statistical modalities. If, for example, flake lengths from a site are measured, we can expect to see clusters representing characteristics from each stage of production.

Some care is needed to account for flakes that were reworked or removed from the site. While scars on preforms can accurately indicate original flake lengths, measurements from quantities of flakes can reconstruct a pattern of original flake inventory.

Since a bell curve indicates the intent to make something fit a standard, the range of variation must tell how successfully the intent was met. Narrow variations might mean that the craftsmen were particularly skillful, but it is more likely that some sort of process-step control forced a standard result.

In our investigations, Tony Baker and I used statistics to address whether the first channel flakes removed from a preform might be thicker than those removed later. Logically, it seemed that because the preform is thicker for the first flute, the channel flake would be thicker. Although channel flakes did not at first appear to retain any indication of which face they originated from, fluted preforms often retain unambiguous evidence of flute scar order. We found that the width of flaking in preparation for fluting to be invariably greater for the first flute than for the second. Armed with this knowledge, 80 Folsom channel flakes found near Albuquerque, N.M. were visually ordered from smallest scar interval to largest. If flutes differed in any significant manner between faces, the characteristics of the lowest 40 should have been different from the highest. As it turned out, the two sets had the same thickness, which indicates that first channel flakes from a thin preform are statistically indistinguishable from those made on a thick preform, even though, statistically, channel width was different between the two faces.

Statistics are not always suitable for resolving how actual knapping behavior was used to control a process. Experimental replication provides the best way to understand how processes can be forced to conform to standards. Even "successful" replication may fail to yield the same spread of data seen archaeologically. The entire production sequence must be replicated to obtain meaningful data matches, it is not enough to mimic an end product.

Summary

Stone can be fractured in such a variety of ways that one process can often be distinguished from another. A forensic analyst must first determine the context of an action, then consider the limits of variation, and finally assess which factors controlled the outcome. Clues are found in archaeological data and anthropological observations, while theoretical modeling and experimental replication reinforce the interpretation. Each time that archaeological evidence is properly interpreted, it adds to the set of case histories that can be drawn on for new studies.

From earliest times, people clearly understood that they could acquire specific flake outlines by setting up the proper core geometry. They also had no trouble with repeating a planned series of flake patterns to control the outcome of their stone tools. We must avoid the trap of thinking most of the debitage people left behind is pure waste. If we can see each flake as serving a purpose of generating the needed core surface, the process of reduction will appear less chaotic and each fragment gains meaning. Piecing together the stories from individual artifacts builds a detailed account of tools, techniques, and strategies favored by a knapping tradition.

By using multiple, independent lines of reasoning, we can greatly improve the chances that the past is correctly interpreted. These examples provide only bare glimpses of what may ultimately be gleamed from careful reflection on remnants from the past. Next, we will take a detailed look at an entire process that has been deciphered through lithic forensics.

Chapter 6

Details of Fluting

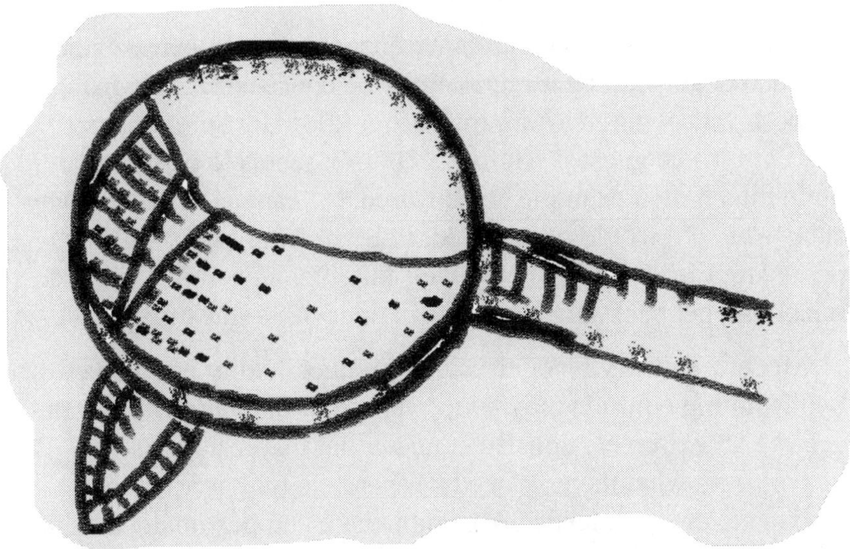

I have encountered the attitude among non-knappers that fluting is no big deal—a flute is just another flake, and there are only so many ways to make flakes. Yet, as we have already seen, fluting calls for the most specific of conditions, and those conditions are not necessarily easy to achieve. Ancient practitioners of fluting worked with an explicit list of process steps, and had methods of using their tools that assured the correct outcome. One might achieve a flute-like flake by accident, but only by carefully crafting a process can such flakes be made routinely. This chapter addresses the explicit sequence of process steps and controls, the "recipe," if you will, that enabled the fluting of Folsom points—the ultimate expression of fluting virtuosity. At the end of each section, variations on these processes are also noted for related technologies such as Clovis, Gainey, Barnes, and Cumberland.

Fluting is made possible by strictly controlling every part of the process to ensure a reliable outcome. By attending to the success of each step, the emerging product nears the limits of what is physically achievable. To look at a Folsom point, the flute seems most important because of how much of the face it consumes. However, the point must be flat and ripple-free to fit its hafting snugly. Controls incorporated into the production process determine each of those qualities. Doing the same thing the same way each time leads to high rates of fluting success.

Each of the process controls we encounter, exist because the tool maker considered it important to be consistent. But what purpose was being met? We may never know for sure, but can make informed guesses. Hafting certainly seems to be important, but until a hafted example is recovered, we cannot be certain why such precision would be required. Experimental use of Folsom points does make it apparent a well-fluted point is much easier to repair and get back into service than one with a thick tip.

Modern industry recognizes that product quality is achieved by instituting controls that reduce variation. Standardization can achieve an extremely consistent result, but that result must be in sync with what the user needs. Where the tool user is the one who made the tool, feedback is immediate and personal. Ancient production was driven by experience to institute controls, which insured that implements would function as intended. Since the concept of process controls is new for archaeology, each of the many sequential steps in making a Folsom fluted projectile point will be reviewed to see how toolmaking can be so consistent.

Process management

The pattern of artifacts left by Folsom people is remarkably uniform at site after site. That leads us to surmise that quality controls were incorporated into each of a progression of simple tasks, producing highly uniform projectile points in spite of many opportunities to fail. Simple, repetitive actions that yield the expected result regardless of an individual's level of dexterity force each step of the process to conform to standards. It appears that most of the controls can be described as straightforward, mechanically oriented procedures. Such process control could have been ingrained by generations of coaching. Alternatively, processes may have been maintained through ritual, as seen in traditional

Japanese archery, where every action is prescribed and controlled and the outcome is inevitably precise.

The Folsom recipe

Once a knapper sets out to flute a preform, work is unlikely to be interrupted until the fluting stage is complete. Thus, a stage is equivalent to an episode of production activity. The rare archaeological examples of work abandoned in mid-stage are usually restricted to discarded failures. Each action uses its own set of controls to assure consistent results and improve the likelihood of success in the next step of production. Jeffery Flenniken (1978) was the first to document how the major phases of reduction were represented in archaeological collections.

Select the right type of stone:

Most Folsom points are made of high grade, slick-surfaced stone with no evident grain or flaws that might make fracture path unpredictable. For the remaining process to have the best chance of success, the stone should have the right balance of stiffness and elasticity.

Experimental replication of fluted points shows that the stones chosen by original practitioners generally can be fluted to specification without alteration. Occasional use of heat-treating to weaken particularly intractable stone is reasonable, but it clearly is not an absolute requirement. Making a stone too fragile by heat-treating could interfere with successful fluting. Obsidian projectiles may be easier to make than those of chert, but they are far more vulnerable to damage. The slick surface observed on many Folsom projectiles may result from superior quality stone or having been stored in water, rather than from heat-treating. The atomic bonds of silica break far more readily when water vapor is present (Michalske and Bunker 1987), and regain their strength when the stone dries. Preforms of tough material may represent stone destined for heat-treating, or such preforms might have been abandoned as unsuitable for making a fluted point because heat-treating was demanded.

Matching hammer stiffness to the preform stiffness may require the worker to adjust the fluting stroke or prepare preforms slightly differently for each type of stone. This potential change of a worker's motor habits should be taken into account when puzzling over distributions of stone far from its source. Since variation interferes with maintaining high quality, Folsom workers may have been reluctant to put new stone types into service until reserves of more tractable stone were exhausted.

> **Clovis**- Clovis points were usually made from slightly tougher grades of stone than Folsom. The largest Clovis points, usually found in caches, often are made of particularly showy stone with distinctive patterns and strong colors. This separates them from ordinary projectile points found in camp or hunting context, which seldom show much concern for aesthetic value of the rock.

Prepare stone for transport:

Unmodified rock is bulky and heavy, so considerable effort at the quarry was required to keep pack loads manageable. At this stage, unusable cortex would be removed and flaws in the stone revealed, creating preforms of shapes and sizes suitable for long-distance travel.

Suitable stone was prepared for transport by applying blows against a platform edge that was offset from the face plane of the core. The offset helps control the mass of the flake and reduces the number of strokes needed to complete the task. It also produces sufficiently thick flakes to be made into projectiles or used as tools, like scrapers. Flakes from this stage will typically retain cortex. These offset platforms might appear to have been prepared carelessly, by battering the edge, but the expert removal of the flake tells otherwise.

> **Clovis**- Enough caches have been found to show that large platters, in excess of 10 to 15-mm thickness, were favored at the earliest stage. The bulkiest platter cores (Width/Thickness ratios below 4:1) have the most rough-hewn appearance. Early control over W/T means that Clovis knappers did not

use dimensional staging, as has sometimes been proposed. Earliest stages seem to show a preference for lengthy flakes that also develop the surface contour. Wide spacing of flakes assured significant utility. Clearly, sizeable flat flakes would have supplied many camp tools. When projectiles were needed, the large platters were broken into sections big enough to be transformed into preforms. Remnants of the squared edge are common, and speak eloquently to the progression from platter, to preform, to projectile. At this stage, it is more important to have ridges that can provide thick flakes than it is to pattern the scars. Platforms on the earliest biface stages are commonly beveled. Often, the platform is no more than an inclined plane facet.

Develop uniform preform thickness:

Prior to fluting, preform thickness was made reasonably uniform in order to facilitate later stages. Occasionally, a flake-blank is flat and thick enough to convert into a fluted point. In this case, the flat surface serves as a control on thickness. Another method of controlling thickness uses widely spaced, flat flakes to flatten the back face of the preform. The third option is to fully biface the preform to uniform thickness. Regardless of the percussion

Folsom biface

This example was composited from the two faces of a broken jasper biface found at the Antelope Spring site in Colorado.

strategy, individual flakes were controlled by specific platform offsets and tool inclinations that led to highly predictable flake geometry. There is no reason to suppose that Folsom workers had any preference beyond convenience or personal habit for the means by which they achieved a flat preform, and each practice is evidenced archaeologically.

Flat flakes are likely when the preform is supported about its perimeter. While overshot flakes are present in Folsom assemblages, they are far less numerous than at Clovis sites. The trend toward flat flake scars makes it easy to build wide, thin knives without having to depart from normal knapping practice.

Most flakes recorded at the Lindenmeier Folsom site are between 15 and 45 mm long, which implies that early-stage preforms were as wide as 45 mm. Overall, bifaces ranged in width from 25 to 47 mm, with most at 34 mm. While recovered bifaces range in thickness between 6 and 14 mm, they probably started between 17 and 55 mm to supply all recovered flake sizes. This agrees with the evaluation from the Stewart's Cattle Guard site that tool flakes must have come from cores in excess of 15 mm thick.

> **Clovis**- Flattening of the first face is quite noticeable at quarries. The second face often develops a dome. Clovis bifacial flakes are typically more curved than Folsom flakes, likely reflecting support at the far edge of the preform.

Straighten the face to be fluted:

Baker's computer simulations of fluting show that the preform face should ideally be straight to control both flatness of the flute and thinness of the finished projectile. Variations in cross-section lead to variations in stiffness that cause ripples in the fracture front, which tend to mirror the surface topography. Surface irregularities of as little as 1 mm can theoretically change the effective stiffness of the preform enough to divert the fracture from its intended path.

By establishing a uniform surface at the percussion stage, less effort is needed during pressure refinement. Relatively few strokes may be required to prepare the preform. Early blows on highly reduced platforms set far from the face plane of the biface allow a few thick flakes to accomplish what it would take with many thin flakes. Fingertip support at the far edge promotes flakes that clean off the preform surface and often reach the opposing edge.

Clovis- Makers of Clovis points tapered the longitudinal profile from approximately midpoint to either end prior to fluting. Early stages of percussion show a strong preference for diagonal percussion flakes, with flake sequences running generally from tip to base. Frequent overshots result from deliberately supporting the preform at the far edge rather than distributing support around the biface perimeter. The highest priority was on making evenly spaced scars extending from edge-to-edge, right-to-left as viewed from the base. Flake scar shape depended mostly on using surface isolation, but also on adhering as much as possible to a sequence.

Barnes and Cumberland- Full-length fluting requires linear ridge-crests free of longitudinal arc to minimize errors.

Refine the preform by pressure:

Supporting the preform on a tabular base of wood or antler improves a knapper's control of platforms. With the preform resting on this support, a narrow rod of antler or bone is applied perpendicular to the tabular base to bevel the preform edge. Because impact is set back from the edge margin, each small wedge-shaped flake is steep and creates a bevel. By gauging against the plane of the tabular base, the bevel can be kept true. Parts of the edge that do not parallel the tabular base are adjusted so the edge falls roughly on a plane. Shaping can be started at this phase as well.

Once the bevel edge is complete and lightly ground, a spatulate antler tool is used to remove a sequential series of flakes from the preform supported in the knapper's palm. After building up considerable compressive force, an outward wrist flick detaches a parallel-sided flake with a distinct initial bulb of force. Adjusting spacing between flakes allows them to just reach the center of the preform. As the preform width declines, so must the interval between pressure flakes. A straightedge of bone or wood helps assure straightness of the preform face but, with experience, sufficient accuracy is possible by eye. A properly completed preform has virtually no surface irregularities, and a shallow trough near the edge isolates the central ridge.

Clovis- Minimal pressure was used to refine Clovis preforms prior to fluting.

Gainey and Barnes- Contouring of faces by pressure is very similar to that used for Folsom points

Midland- Midland points do not retain flute scars, but some archaeologists believe they are made by the same people who made Folsom point. The specialized flaking from a beveled platform reached exquisite perfection in the best Midland projectile points. Flakes are broadly and uniformly spaced, even when they are not sequential, and often reach to the far edge.

Goshen and Plainview- While these styles were not intended to be fluted, the style of pressure flaking and the surface contour are comparable to Folsom technology in both approach and execution.

Prepare preform ends:

Beveling to define a fluting platform uses the same tabular chipping base control as used to prepare for pressure refinement. The fluting nipple needs to be only slightly isolated, and ridges between flake scars often create a shallow buttress at the back of the nipple. Leaving the nipple offset from the face plane by 1.5 to 2 mm promotes optimum flake development. Archaeological preforms show slight sanding along the leading edge of the bevel, perhaps to avoid snagging the edge during fluting. After the nipple is ground with a bias to the face, it can be polished with a piece of debitage.

The tip is blunted while using the tabular chipping base, to position contact with the anvil near the center plane and offset at least 2 mm from the face plane. Sometimes the blunting amounts to beveling, but it often has no particular structure. Thick tips add

Folsom fluting platform

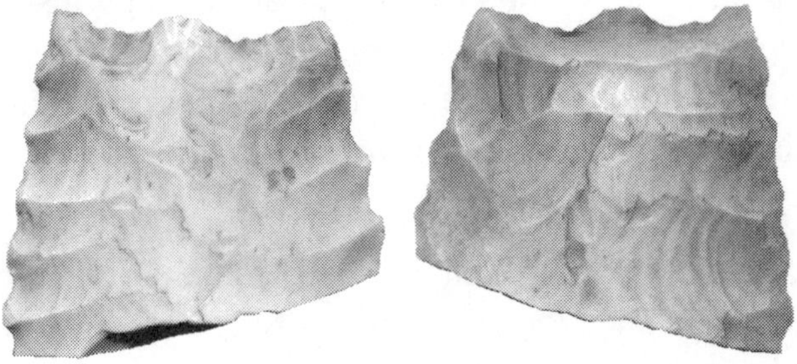

dorsal ventral

Front and back views of a broken Folsom preform from the Bobtail Wolf site near Lake Ilo in North Dakota, show the details of platform construction. Strong isolation of the platform gives it the look of a nipple, as it is often referred to. The back side is beveled to a plane and then the dorsal ridge is structured to facilitate easy development of the flute. Also note the even spacing of flakes that define the dorsal ridge.

enough stiffness to cause a step termination in the channel flake, just short of the tip. Width of tips can vary from about 1 cm to over 2 cm. Heavy grinding at the tip completes preparation of the preform for fluting and controls against the tendency for tip damage on an anvil. Thin experimental preform tips shattered against a wood anvil and sometimes split it. Thick tips help avoid those problems.

Clovis- Earliest Clovis forms were fluted when the tip was sharp because full-length flakes were not the goal, and anvils were not used.

Gainey- Beveling of tips has been reported.

Barnes- Tips were blunted and sometimes squared, but apparently not beveled.

Flute the first face:

Archaeological evidence points to percussion as the preferred method of fluting. Other methods will be detailed in chapter 10. Supporting the preform near the tip expands the 'sweet spot' of conditions suitable for fluting. The tip needs to be held in tight contact against an anvil of bone or wood to avoid end shock from bouncing. The anvil at the tip effectively adds enough resistance for a full-length flake to develop. If the flake runs out of energy, it can terminate in a step far before it nears the tip. Too stiff an anvil can cause hinge terminations that shorten the flute, and too much energy tends to make the flute erratic.

It is not easy to precisely control a percussion blow. Using hammer mass to balance the anvil resistance can be quite unwieldy, especially when taking a long stroke. Subtle control is possible by flipping a heavy baton with the wrist, and using angular velocity to effectively transfer energy with a short stroke. By contacting the platform with the baton at a slight angle to the preform, even a large diameter billet can reach into the basal concavity without damaging the preform ears. The strategy chosen for the blow should have more to do with what the knapper is comfortable with than what is deemed "right."

As in every other percussion stroke, the baton has to be inclined to balance between outward and longitudinal forces, and the blow angle has to be low as well—not an easy thing to do. Since it can be very difficult to deliver a percussion stroke with a predictable orientation at the point of contact, the knapper needs to find a body position that controls the angle of incidence. For example, holding the preform close to the knapper's body tends to force a tool position that promotes full-length flakes that propagate straight and flat.

Clovis- The first flake was positioned to remove the greatest thickness and can appear to be misaligned. Additional fluting is likely on the completed projectile. Feathered terminations are most common for Clovis flutes since most flutes were not intended to reach full-length. The angles of blow typical for feathered flakes will not lead to full-length flakes unless an anvil is used.

Remove excess thickness from the fluted face:

Even if the first flute is narrow and shallow, it can serve as an effective control for the remaining process. The key is in achieving a uniform surface so that subsequent fluting attempts on that face have minimal reason to ripple. For the same reason, terminations remaining from incomplete flutes need to be re-flaked to avoid the bending stress concentrations that abrupt changes in thickness promote.

> **Clovis**- Excess thickness and step terminations from fluting were normally removed by percussion rather than pressure. Pressure refinement usually obliterates evidence that preforms were prepared one face at a time.

Refine the second preform face by pressure:

Using the tabular chipping stage again, beveling for the second face is easier than for the first because the preform has already been brought to a plane by percussion and relatively fewer surface imperfections should remain. Correcting major surface impediments at this juncture risks snapping the preform and, in fact, many archaeological examples confirm that preforms were broken while preparing the second face for fluting. At least one preform, recovered by Dr. Dennis Stanford from the Zapata site in Colorado (2003), shows that pressure flakes were taken from a steeply beveled perimeter, one edge at a time.

> **Clovis**- While Clovis points were often made one face at a time, examples occur that indicate that sometimes the refinement was applied equally to both faces. Pressure techniques are not as evident as percussion.

Prepare preform ends:

The preform ends need to be prepared in the same way as on the first face, except that now the preform is more delicate. At this stage, the preform ears present a greater psychological barrier, since the blow has to clear them by only millimeters. Close inspection of archaeological preforms at this stage shows that ears have invariably been shortened by retouch because the beveling always extends the full width of the base. Completed projectile points may give the illusion of having ears too tall to accommodate fluting by direct percussion, but they usually have retouch that removes the nipple remnant, deepens the concavity, and emphasizes the delicate nature of the ears. Finally, the relationship of the tip contact with the center plane has to be checked and adjusted to minimize bending stress.

> **Clovis**- No tip readjustment was needed for Clovis fluting since the anvil effect was often satisfied by no more than finger support. The fluting nipple was set near enough to the center plane that no further adjustment to the basal taper would be needed after the flute was removed.

Flute the second face:

The blow to remove the second channel flake needs to be as aggressive as for the first removal. A weak flute makes it difficult to complete the taper to the tip by pressure flaking. The flute is successful if it extends nearly to the tip, regardless of how it terminates. Finished Folsom points leave the impression that the final fluting blow reaches impossibly deep into the basal concavity. While some blows undoubtedly did reach significantly between the ears, the deep concavities are usually developed by retouch after the flute has been removed. Ear damage on late-stage failures is not common, so we can assume that familiarity with the process developed enough accuracy that errant strikes were rare. However, split preforms and burinated edges are relatively common.

> **Clovis**- Fluting the second face was relatively simple for Clovis points since the basal concavity was normally shallow.

Evaluate thickness, and re-flute if needed:

Archaeological examples of multiple fluting removals on the same face are fairly common. As noted earlier, deliberately shallow flutes on the first attempt can serve as a control to reduce bending on the preform. Many points were re-fluted, even when they were already as thin as other projectile points that had been put into service. While there may have been functional reasons for continued thinning, it undoubtedly adds to the risk of failure. Repeated thinning by fluting assures us that Folsom knappers had confidence that their process would succeed even under extreme conditions.

Preform details

Above: Two replica preforms at actual size show typical flake scar spacing for the first face. Spacing gets closer as the preform is narrowed Also note the blunt tips, bowed sides, and relatively narrow base.

Below: Portions of Folsom preforms from the Antelope Spring site. The split preform shows that pressure flaking on the unfluted face has closer spacing diagnostic of second faces.

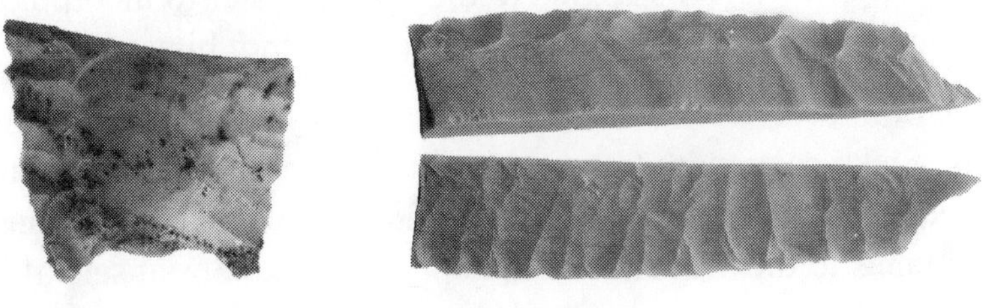

Clovis- Multiple fluting was as common on Clovis points as on Folsom. One small projectile point from the Mockingbird Gap Clovis site in New Mexico appeared to have been fluted at least nine times before being discarded (Weber 2001).

Gainey- Second fluting on a face is known, but the final flutes stop just short of the first flute termination. The second flute seems to be used to taper the base.

Barnes- Fluting was not repeated on a face, although a special thinning flake was used on one or both faces to thin the base for hafting.

Remove excess thickness between the flute termination and the tip:

Tapering a preform tip requires a style of pressure flaking different from the earlier stages of production. Replication shows that the preform has to be firmly supported on edge to let the flake shear away convexity beyond the flute tip. A few widely spaced flakes leave a uniform pattern of flat flake scars traveling well past the centerline. A spatulate pressor used on edge creates flakes typical of those seen on tips of Folsom points.

Clovis- While step flakes can sometimes be observed at the end of Clovis flute scars, a lateral flake was commonly taken after fluting to leave a smooth transition from the flute scar to the remaining surface profile. The Fenn Clovis cache (Frison and Bradley 1999) shows that the highest priority was given to evenly spaced flake scars extending diagonally from edge-to-edge, right-to-left as viewed from the base. Consistent flake orientation would provide the best blanks for tool flakes, while maintaining the surface profile. Twice as many major flakes were oriented toward the tip as toward the base, although some individual faces show a flaking bias toward the base. Overshots angled to the tip are easier to incorporate into a projectile shape than when an overshot bites away extra width near the base.

Flake sequences trend generally from tip-to-base, although sequences from base-to-tip are used on occasion, sometimes counter to the sequence of the opposite face. Serial sets of two to eight flakes each were interrupted when the knapper recognized that making the next flake would leave an undesirable scar because there was no guiding ridge.

Remnants of the stage prior to the final patterned sequence indicates that flakes were carefully positioned to allow maximum travel. Each series of flakes was also designed to improve on the visual effect of the prior stage. Once final projectile thickness was attained though, long flakes were discontinued. More thinning would interfere with hafting, and weaken the point. Strongest patterning is on the *thickest* projectiles, possibly because there was ample thickness to control the flake pattern.

Refine the edge:

The tabular chipping base is essential for controlling the final refinement of a Folsom point. Applying the slender retoucher in the same manner as was used to prepare bevels for pressure refinement allows very clean, small flake scars to be placed in close sequence. The rigidity of the chipping base assures that the retouch flakes are straight and do not arc into the channel flake scar. A few Folsom points lost at this stage show that when a wide perimeter needed to be reduced, the beveling began with widely-spaced flakes that left a serrate edge. The projectile edge appears to be turned into a saw, but it is more likely an expedient method of getting to the final bevel more quickly.

Repeated series of flakes can continually improve the quality of the retouch. Axial load, applied by a thin, pointed tip of antler or bone, accommodates retouch spacing of as little as 1 mm. Pronounced grooves on sandstone slabs at the Lindenmeier Folsom site may represent the shaping and maintenance of such pointed knapping implements. Edge abrasion suitable for earlier stages of work would make edges too strong for retouch, so polishing with a piece of debitage is usually just right.

Clovis- Percussion techniques were used for much of the edge refinement, especially on the largest varieties. Re-used points tend to show a great deal more work by pressure techniques.

Barnes- Edges were greatly modified by selective pressure, so earlier stages of contouring are not readily visible.

Adjust the shape:

Although shape becomes more evident as work progresses, the final nuances are added by sanding. A gentle, grazing stroke with a gritty sandstone snags and removes small projections along the edge. Attrition can be controlled with enough accuracy that subtle adjustments of shape are possible. It is likely that the desired shape came from a mental template imbedded by a lifetime of acculturation. Part of that template undoubtedly involved the hafting requirements for Folsom projectile points.

Retouch

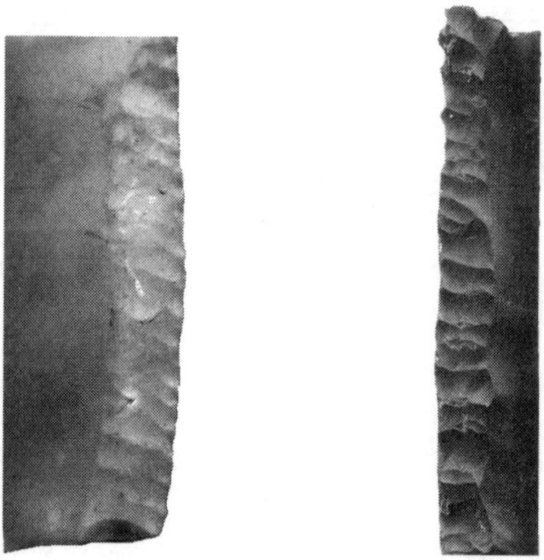

Two examples of replica retouch are shown at twice their actual size. The scars terminate cleanly without intruding into the channel scar. Sequential ordering develops uniformly parallel scars.

Clovis- Clovis knappers were especially adept at building symmetry and balance into their projectile points with a minimum of flakes, and not overworking the edges—especially in the first incarnation. Detailed shaping appears only as maintenance is necessary. The wide initiation, short travel, and feathered terminations, not to mention the lack of pressure nibbling indicate that retouch was likely done by percussion. Pressure retouch would leave more small steps than Clovis artifacts usually exhibit. During experimental replication, a large, gritty hammerstone produced equivalent retouch scars when swung by slow sweeps to strip flakes from the lower surface.

Gainey- Early stages were thoroughly reworked by taking short flakes after the flutes were removed.

Barnes- Bases were narrowed and given a fishtail shape.

Smooth the basal edges:

Once the shape has been established, the sanding action is extended to thoroughly dull the basal edge. While edge grinding is well recognized, additional polish was applied to the ground edge with a flake of debitage. The polish may be an aid for seating the projectile in a standard haft, but it also removes stress points that could lead to breakage during use.

Clovis- Much more aggressive edge grinding was used than on Folsom points, suggesting that Clovis points were subjected to greater stress in use. The same tool used laterally to sand the edges can be used longitudinally to smooth the edge.

Dimensional Data

Many anomalous point "types" actually represent modifications to the original projectile form during a life cycle of damage and repair. The thinnest and shortest points probably survive as small versions through salvage and retouch rather than because they started small. Reducing length and width of unused experimental replications makes them agree with median archaeological values.

Archaeological reports typically relate dimensional data as a *mean value* (X) of a given *number of specimens* (N) and *standard deviation* (s). Generally, we think of data being evenly distributed in the manner of a bell-shaped curve, and 95% of the data is within two standard deviations of the mean. When the data is skewed, as seen in length measurements, it alerts us to look for causes. In the case of length, the cause most likely stems from damage and repair. Since we are interested in comparing the effects of processes on various attributes, dividing the standard deviation by the mean produces the dimensionless *coefficient of variation* (CV), which normalizes unrelated measurements for comparison without regard to scale. The lower the CV, the less variation is present.

Folsom artifacts have been shown to manifest a remarkably consistent CV over a wide geographical area, regardless of which dimension is used. For example, Folsom CV values are 0.06 to 0.16 (Tunnell and Johnson 1991). Even attributes like length, that are extensively modified by damage and repair, show low variation. We conclude from this that, despite geography, the Folsom response to environment manifested in a uniform way. In Clovis, the same attributes show considerable variability. Clearly, Clovis processes were not controlled nearly as stringently as Folsom.

A scattergram of Folsom length and thickness data from many sources illustrates some of the complexity of interpreting dimensions. Increased thickness for long preforms shows the limits imposed by physics, especially on the thin side. Upper thickness limits are at least partly a matter of choice, based on hafting and social decisions. Experiments confirm that it is difficult to flute a replica thinner than original specimens. At the upper end of known preform lengths (100 mm), it is difficult to keep the maximum thickness within the limits tolerated for shorter Folsom projectile points. Although length obviously varies greatly because of rejuvenation, the mean length of archaeological specimens is remarkably consistent from any part of the Folsom range of occupation. This indicates a highly consistent Folsom era lifestyle that reveals the same pattern whether in the North Dakota badlands or the South Texas hill country. With the foregoing examples in

Folsom data

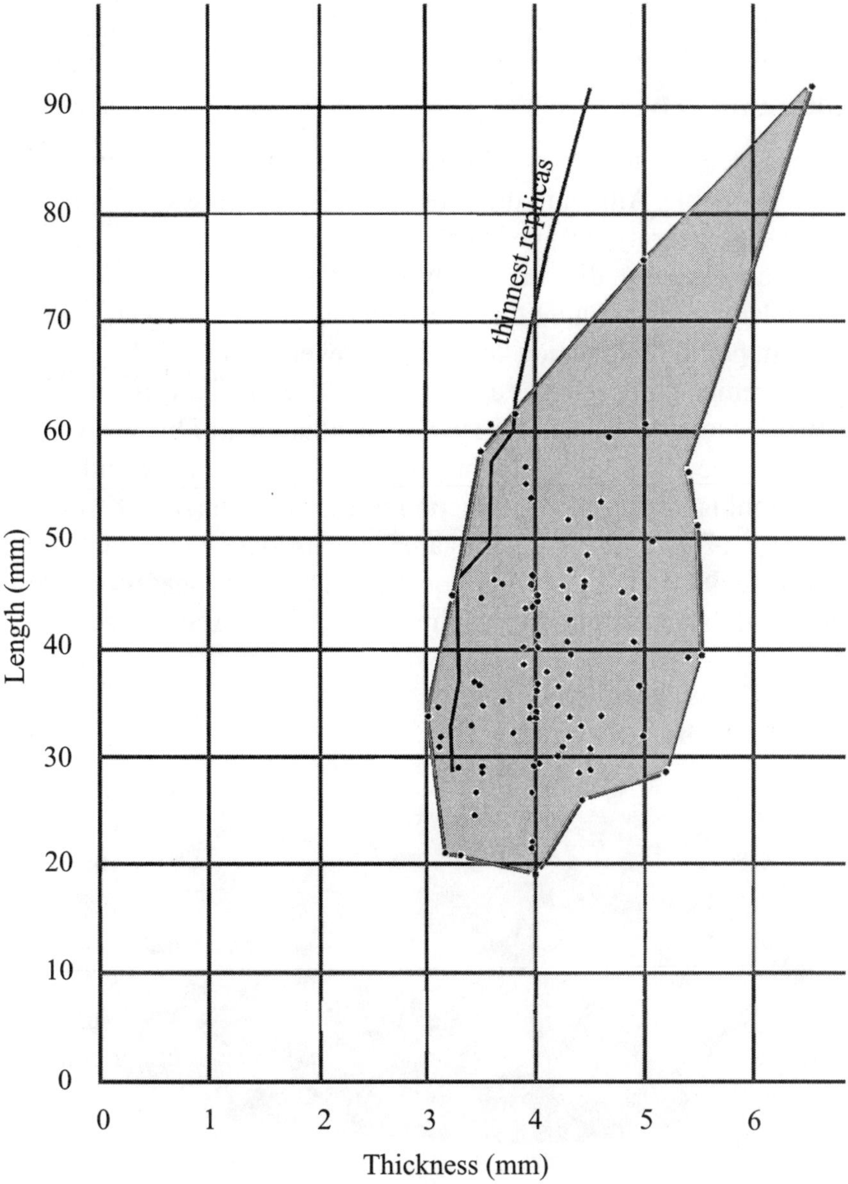

Shaded area represents dimensions of complete Folsom points from many sources (Largent, 1991:333-334; LeTourneau,1997). The longest points tend to be thicker because of increased risk of breakage in manufacture and in use. Thinnest replicas approach the same physical limits encountered by early knappers of Folsom points.

mind, I offer a hypothetical "envelope" of acceptable variation in Folsom projectile dimension, represented by the shaded area in the illustration on page 155.

Ultra-thin knives

A separate strategy was employed for special knives used by Folsom people. Although they are actually present in many sites, these knives rarely survive in easily recognizable form. Called ultra-thins because they are nearly as thin as projectile points but much larger, they require the same technology used in preparing Folsom points. Production strategy, however, is quite different. Measurements of preforms excavated at the Bobtail Wolf site, South Dakota (William et al. 1997), reveal that quarry blanks were selected specifically for their breadth and initial thinness, and no other tool is on the same production track. From the outset, until the knife is ready for use, flakes are made as long, wide, and flat as can be controlled. As a consequence, at the later phases of thinning, flakes start to dive slightly at the center of the biface,

Ultrathin knife

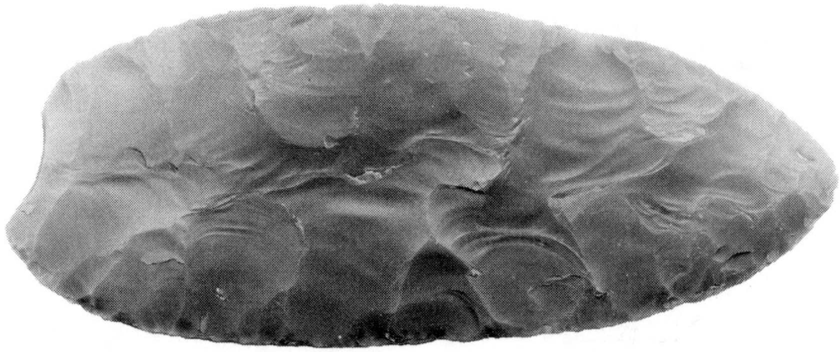

This replica of an ultra-thin knife is only 5.4 mm thick, although it is 120 mm long and 48 mm wide. Since almost all ultra-thin knives are broken or reworked, the shape is conjectural. Sharpening flakes on this knife style tend to create beveled edges. Similar knives were used by most hunters of bison.

Reduction sequence (dimensional ranges annotated in mm)

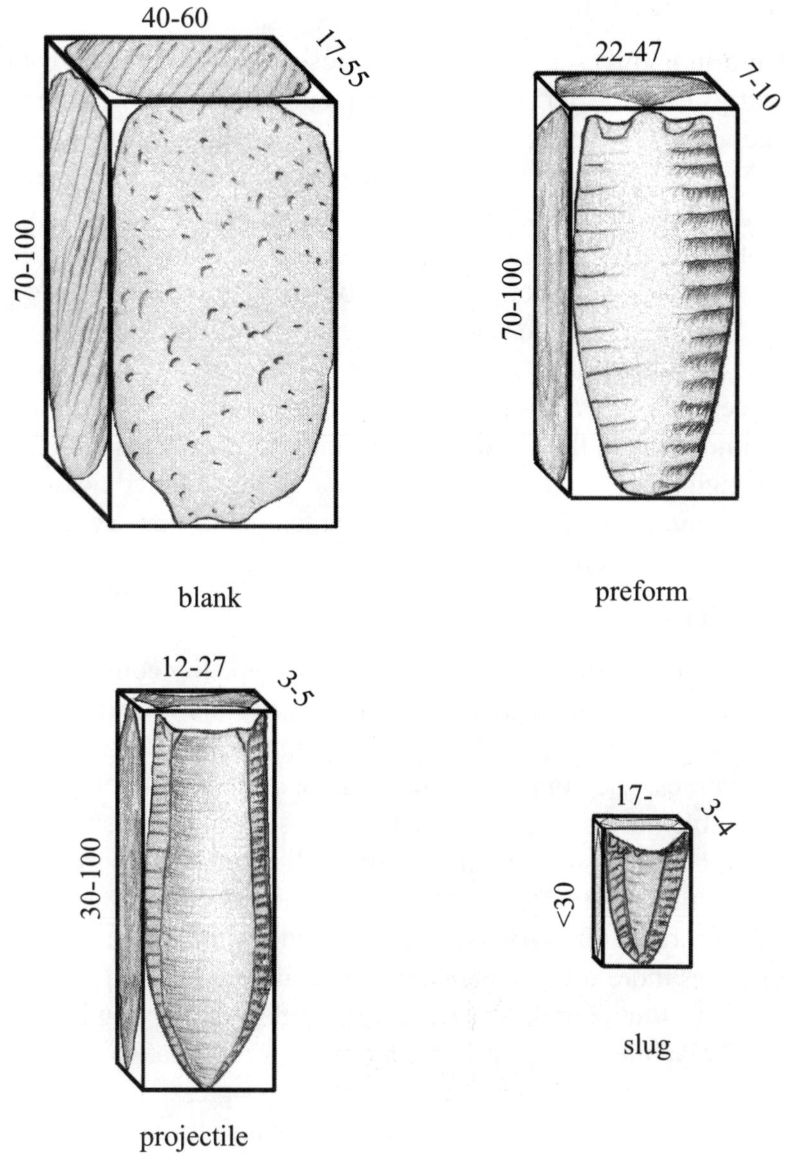

Progressing from initial quarry blank to discarded slug, Folsom knappers tended to stay within prescribed size ranges.

often leaving the center thinner than the remainder of the tool. Only short beveling flakes are then necessary for most of the remaining life of the knife.

Ultra-thin knives are thought to represent tools for slicing meat into strips to dry as jerky. It is easier to cut thin slices when the tool reaches easily into the cutting area. Since ultra-thins are too thin for other purposes, and too risky to make while on the trail, they appear to have been made primarily at the quarry. If they had to be replaced away from the quarry, it may have compromised the balance of flake tools normally available from the chosen blank. Upon breaking the knife, segments were typically pressed into service for other tasks. Quite often the knives were broken into radial segments by a sharp blow at the center. Flat flakes were sometimes broken the same way to produce unique tools with sturdy right angles, with acute tips. Some of the unfluted Folsom points, about 25% of the total, have characteristics that suggest they originated from broken ultra-thins.

Conclusions

Returning to the formal practice of industrial quality control that we touched on at the beginning of this chapter, an even stronger correlation can now be made. Testing equipment, published specifications, and computer databases are merely more precise and tangible means of control than human senses, experience, and memory. Items that are made repeatedly, with any degree of uniformity, must adhere to the same basic principles. Just as industry recognizes some equipment and manufacturing practices as more reliable than others, we must concede that people, substituted into the same roles, perform at varying rates of reliability. While testing concepts by experimental knapping during the writing of this book, my results were of progressively higher quality. The improvement was primarily because the use of process controls allowed me to avoid the small errors that otherwise would have forced me to improvise a unique solution. In effect, my replications became clones of one another because variation was squeezed out of the process. Early knappers undoubtedly reaped the same benefits from using standard ways of making everyday tools.

Statistical comparisons of any product reveal the overall rate of success by ranges of variation. Especially when considering the physical tools available, Folsom production appears to have reached the most impressive levels of control seen in fluting technology. Subsequent lithic technologies seem to generally decline in uniformity of product as time progresses.

People made their tools the same way each time because the tools had to be dependably available. The mix of various tool types is so consistent for each culture that it becomes a way to distinguish one group from another. For us to understand the underlying causes of differences in how tools are manufactured by successive cultures, we will need detailed descriptions of each technology being compared.

Several obstacles make the task difficult. A large volume of data must be recorded and compared, merely to ascertain how well a process has been controlled. Proposed methods of construction must be actualized by experiments and the results validated by lithic forensics. It is equally important to study how people dealt with the inevitable limitations of resources and the demands of their environment. The next chapter shows how a lithic technology can satisfy every other stone tool need as a by-product of making projectiles.

Chapter 7

Life Cycle of Stone Tools

Whatever the purpose of an implement, it must be viewed as having a life cycle spanning its conception, creation, use, damage, repair, and abandonment or destruction. Otherwise, we can too easily lose track of the phase that is represented by evidence recovered archaeologically. If the tool maker's plan were always carried out, there would be nothing but debitage and exhausted implements to consider. Fortunately, we can rely on the unforeseen to fill in the details of what typically happened to each category of tool. Miscalculations and misadventures of past people provide a treasure trove of data that archaeologists can use to reconstruct ancient lifeways.

For nomadic peoples, there are numerous limitations as to how much load can be carried, how much waste can be tolerated, what use functions must be met, where raw materials can be obtained, and much more. While each group learned to accommodate these constraints, the means of accommodation is not always readily recognized. This chapter will develop some concepts that enable rough estimates of the magnitude of resources needed for various groups to maintain their lifeways. Even if exact quantities cannot be specified, relative differences in strategies can help us to understand why certain coping mechanisms were adopted and what their consequences were.

Stone tools progress through a simple cycle of acquisition, construction, use, and discard. Of course stone must be acquired at the start, but the rate of input varies with how easy it can be gathered and processed. Once stone goes into use, there are good reasons to maintain relatively steady inventories of partially completed work at each stage of the process. Eventually, in spite of tool maintenance and repair, stone is disposed of by a combination of discard, loss, or trade. Whether a quarry, base camp, or kill location, the context of a site illustrates considerably different aspects of the cycle. By combining archaeological data from various contexts with replication results, the economy of stone use becomes reasonably clear.

Lithic materials were continuously moved through their life-cycle. Once stone was previewed and roughed out at the source, it was transported as a biface that would be reduced by increments as there was need for tool flakes. We know how much stone is left at a site, but must use some detective work to figure out how many people were there to use it. We can make some estimates based on the relative amounts of each category of flake or tool. Future sites or new analytical methods may convey sufficient resolution about how many people may have left behind a given quantity of stone.

Acquisition

Prehistoric people literally went to great lengths to acquire most of their tool stone from quarries. Several factors warranted the extra travel to a given quarry. It is necessary to have significant quantities of rock to allow adequate preview and high-grading before setting out on the next hunting round. Quarry stone provides a known, predictable quality. Still, it is not as easy as walking up and scooping a load of rock. Rock at the surface is often flawed by exposure to heat and cold, as well as deterioration due to water leaching silica from it. Buried rock is invariably of higher quality because it is hydrated. Sources are limited in number and are often great distances apart. Deliberate, scheduled quarry trips allowed a group to process and high-grade a significant amount of stone, assuring that what they carried away was of sufficient quality and optimum configuration.

The R-6 workshop site from the Cody complex, excavated near Sapello, New Mexico by Dr. Dennis Stanford (Stanford and Patten 1984), gives us some clues to what activities were conducted near a quarry. Many worn-out scrapers found among debris initially made it look as if hides were processed there. While there were numerous overshot scars found on preforms, hardly any overshot flakes or scrapers made of local stone remained. Overshot flakes are typically uniformly thick and curved, ideal for scrapers. The most reasonable explanation for the abundance of scrapers is that they were normally saved until replacements could be collected from the workshop. Upon replacement, old scrapers would be jettisoned. After leaving the quarry, small tool flakes could have been provided as needed from preforms in progress, but heavy-duty flakes seem to have been harvested during quarry visits. Earlier people may not have adhered to the same pattern of accumulating tool flakes before they were needed. There is reason to believe that Folsom knappers did as little initial reduction at the quarry as possible, and made camp tools only as the demand arose.

It was risky to rely on finding stone whenever the need arose because stone was central to survival. Other sources of stone would not be spurned, but dealt with as opportunity allowed. Rough stone sometimes appears in camp tools, even though it is not good for fluting. Scattered deposits of high-grade rock were too labor intensive to look for where it was not reliably found. When trends of stone used for different tool categories are examined, it appears that most raw material collected during other than scheduled quarry visits was used for tool stone more than for projectiles. Subsequent use patterns by later people show increasingly expedient use of local stone. Presumably the pattern of efficient reduction lessened over time as people became more settled and mobility was less of a driver.

The largest Folsom campsite in the area near Albuquerque, New Mexico, is located near a prominent geologic feature that is visible for miles. Highly visible landmarks would have made many of the largest paleo-period camps easy to locate, even for those who had not been there before. That facilitates gatherings between normally separate bands of people, where social exchanges and re-distribution of resources could take place.

One way to understand how a particular stone gets disseminated away from a quarry is to plot relative concentrations found over a large area as a density map. In the Albuquerque basin, Chuska chert, which originates over 100 miles to the northwest, is most concentrated at the main campsite and becomes more scarce farther out. Each site in the general area has similar ratios of stone types from various quarry sources. Either stone was brought to a central marshalling area and distributed to sub-band members or each sub-band may have simply shared resources as they returned to the rendezvous point.

Other studies indicate that in areas north of the Alibates flint quarry, in the Texas panhandle, Edwards formation flint from the Texas hill-country was distributed more thoroughly than was Alibates. At least three explanations can be postulated to account for this: First, Edwards formation flint occurs over a wider area and represents travel from a variety of sources, while Alibates has to come from a limited locale north and west of the Edwards formations. Second, the direction of seasonal hunting movements caused more Edwards flint to be left north of the Alibates quarry. Third, the working characteristics of each stone type were different enough to cause people to want to use up remaining Edwards flint before working on the Alibates.

To estimate how many pieces of stone—destined to supply camp needs for an extended period of time—that were reduced at a site, we have to be able to distinguish one rock from another. Occasional samples of stone exotic to an area show that trade played at least some part in moving material from place to place. Other varieties of rock are not as readily sourced, but clues from colorants, texture, inclusions, patterning, and ultraviolet response can be used to track their movement with reasonable confidence. Alibates flint, for example, with its distinctive coloration and patterning, has been found in small amounts at sites as far north as Canada.

The largest volume of stone at the Antelope Spring site in the South Park of Colorado comes from a chalcedony outcrop at the edge of the site. Most flakes from the site are relatively thick, and appear to have been made to remove ridges and square edges of raw quarry blanks. Relatively little of the local stone appears as final stage projectiles or tools. Several large pieces were abandoned when broken quarry blanks clearly could not be salvaged as anything but very short projectile points.

In contrast, jasper quarried from at least 10 miles away appears mostly as thin, scalar flakes. Some jasper preforms were abandoned when they broke at a rather advanced stage of preform construction, and only a few jasper flakes retain cortex or show other indication of an early phase of reduction. Freshly quarried stone was reduced only enough to make it easy to carry, while preforms that had already been stripped of their large flakes were turned into projectile points.

The sampled flakes at Antelope Spring demonstrate that about 30 jasper performs were finished as projectiles at the same site where 30 new chalcedony bifaces were prepared for transport. The correlation suggests that final conversion to finished projectile point was deliberately delayed until replacement stone was securely in hand. We may presume that as the group moved away from a quarry, the supply of stone was deposited in increments at one site after another. Bifaces recovered from the Lindenmeier site (Wilmsen and Roberts 1984) in northern Colorado, reflect a similar pattern, with early stage preforms made primarily from nearby stone and late-stage bifaces from distant sources. If we hope to see the full cycle of stone reduction at any one site, that site must have been occupied for a long time over all seasons.

A modern sample of raw jasper, equivalent in quality to that used at the Antelope Spring site, was collected about 15 miles away. Two unused quarry blanks left at the site are similar in size to those in the experimental sampling, and show that 3 to 5 unaltered quarry blanks per pound is a reasonable starting estimate. If filled with 30 preforms as collected at the quarry, a 15 x 13-inch canvas bag like you carry groceries in would weigh about eight pounds. It is likely that en route to the site, some excess material may have been removed and discarded. However, a full bag helps us visualize a pack load and understand how much utility can accrue from such a load. Of course, this analysis is not absolutely precise. The ratios change depending on how far the group expects to travel, what activities must be performed each day, and on how well an archaeologist records data. Overall, though, patterns emerge due to the persistence of a consistent lifestyle. By the time our hypothetical group reached Antelope Spring, many of the blanks had been already been made into fluteable preforms. Out of a potential eight pounds of jasper, only a fraction remained. The rest would have been left as debitage at other sites, some of it having served as camp tools before being discarded.

Production line

Many obstacles stand in the way of estimating how much rock was needed to supply the cycle of lithic material as it progressed from quarry to discard. Weighing recovered flakes cannot account for the inevitable fine shatter, retouch, and erosion. We need a way to estimate the parts that a rock is sectioned into, even when the pieces have been scattered irretrievably. As we have seen, the thickness of a flake is limited by the size of the piece it was struck from. No matter how much a flake is used, broken or otherwise modified, it seldom loses its original thickness, and this allows us to infer the original core thickness to be at least 2.5 times the flake thickness.

To detect a pattern of what it takes to supply a group with the necessary tools to conduct everyday living we must look at the site as a whole. One rock might be barely large enough to be turned into a projectile point, while another rock yields plenty of scrapers before becoming a projectile. The highly variable nature of stones at a quarry forces us to envision an average size of stone removed from it. Even when flaws or mistakes make a rock unsuitable as a projectile, much of the remaining debitage can still be used as tools.

The Lindenmeier Folsom site is the only published site with adequate data to represent a normal production sequence over many seasons (Wilmsen and Roberts 1984). Out of a conservative 51,858 pieces of flaked stone that were collected from a half-inch screen, 0.6% were bifaces, 0.6% were projectiles, 0.6% were preforms, 4% were modified into tools, and 94.2% were unmodified. Since the site was occupied repeatedly by large bands over extended periods of time, the deposition of artifacts should be sufficient to represent the normal movement of stone through a production sequence. In order to understand how camp tools relate to weaponry, though, we need to be more definitive about how portions of the reduction process serve each requirement.

Intuitively, flake dimensions should serve as a proxy for the purpose of estimating the volume of stone reduced at a site. Regardless of stone or knapping tool, flakes from a given surface geometry should differ only by size. I compiled a database of a hundred flake dimensions from Folsom sites, supplemented by representative replication flakes, and found that the true volume of the flakes matched their "boxed" volume of length times width times thickness at a 96% confidence level when multiplied by 0.256. If you only know the thickness, you can approximate the volume by 40.721 times the thickness raised to the power of 2.2578 at an 88% confidence level. Further investigation showed that the relationships best represent sharp-edged flakes, while squared edges can cause flakes to approach their boxed volume.

To test the flake-geometry equations, I replicated two preforms by removing 69 flakes. The replications assured 100% recovery, and allowed me to specify the starting dimensions accurately. By calculating volume from 85 flake fragments, I came up with an estimate of 68.5 flakes, remarkably near the original number. Despite unavoidable approximations due to irregular shapes of quarry blanks, and highly varied flake shapes, the projection is remarkably consistent with the physical count and suitable for analysis of archaeological data.

Flakes per rock by counting

Site	*flakes [implement]*	*projectiles* +	*bifaces* +	*preforms* =	*rocks*	*flakes/rock [impl/rock]*
Lindenmeier	51,858 [2,945]	322	303	323	948	54.7 [3.11]
Shifting Sands	4,618 [258]	92	8	20	122	37.85 [2.11]
Stewart's Cattle Guard	1,131 [610]	104	11	80	195	5.82 [3.12]
Hanson	--- [508]	25	28	27	78	--- [5.56]

Flakes must be large enough to be recovered by 12 mm screen.

This analysis depends on good archaeological recovery, on properly identifying the artifact class, and on interpreting the number of items to be counted. Additionally, the number of flakes produced from a rock may vary according to the rock's size and shape, as well as the knapper's skill. Still, it provides us an approximate estimate of the amount of lithic reduction at a Folsom site.

Experimental volume determination

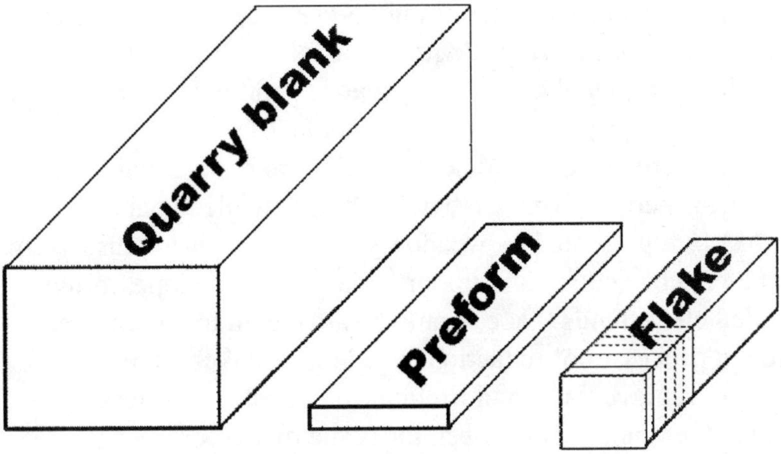

Volume	portion	thickness-range	T	W	L	#
45,000	quarry		15	60	115	
25,175	preform		9.5	53	115	
169.5	6/85	1-2	1.73			4.8
1465.6	22/85	2-3	2.53			17.7
3398.3	26/85	3-4	3.41			20.9
3737.4	16/85	4-5	4.41			12.9
2223.3	6/85	5-6	5.41			4.8
3966.9	7/85	6-7	6.53			5.6
942.4	1/85	7-8	8.18			0.8
1939.1	1/85	11	11			0.8
volumetric estimate						**68.4**
actual count of flakes removed (after refit)						**69**

Histograms of flake measurements from experimental biface reduction of two preforms allowed reconstruction of the flake per rock ratio at 68.4, compared to 85 flake segments physically recovered from 69 flake removals. Trials show that about 10% is unrecoverable.
These results encourage the use of archaeological histogram data to estimate flake per rock ratios at sites.

So few site reports contain detail comparable to that reported from the Lindenmeier site, that I was forced to develop an independent data set by measuring 360 surface-collected flakes from the Antelope Spring site near Hartsel, Colorado. This collection consists of chalcedony quarried and processed at the site, as well as jasper that originated about 10 miles away. No other materials were present, and no other cultural contamination was evident. Jasper preforms were being fluted, while the chalcedony was barely represented by weaponry. Flake-dimension histograms from local chalcedony and imported jasper at Antelope Spring revealed clear trends that can only be attributed to distinctively different processes of reduction. The largest flakes were taken mostly to remove flaws and irregularities. Flake thickness is predicted by the offset between the plane of the core surface and the plane of the blow. As each flake is removed, there is less opportunity to remove another flake of comparable thickness. Successive flake removals also flatten arrises that are required to guide long flakes suitable for end scrapers. Large offsets indicate that starting quarry-rock sizes sometimes exceeded 50 mm thick. Off-margin impacts were responsible for further reducing preforms to between 24 and 18 mm, while 6 to 8-mm flakes from this stage were used for most tools. After that stage, marginal biface work prevailed until the preform reached 9 or 7-mm thickness.

Histogram plots from Antelope Spring identify truncations of frequency spikes between 6.5 and 7.5 mm in thickness, that are probably caused by removing flakes to be used as tools. Tool histograms from the Lindenmeier and Cattle Guard sites show how tools would account for the missing data. The removal of flakes for tools is also visible on the thickness histogram, which shows that bifacing started immediately after tool requirements had been met.

Because lithic reduction processes can differ from site to site, it is necessary to use site-specific thickness histograms for volumetric analysis, but histograms from Lindenmeier and Antelope Spring are similarly patterned. Assumptions of core size were guided by direct evidence from discarded cores where

Flake thickness histogram from Antelope Spring

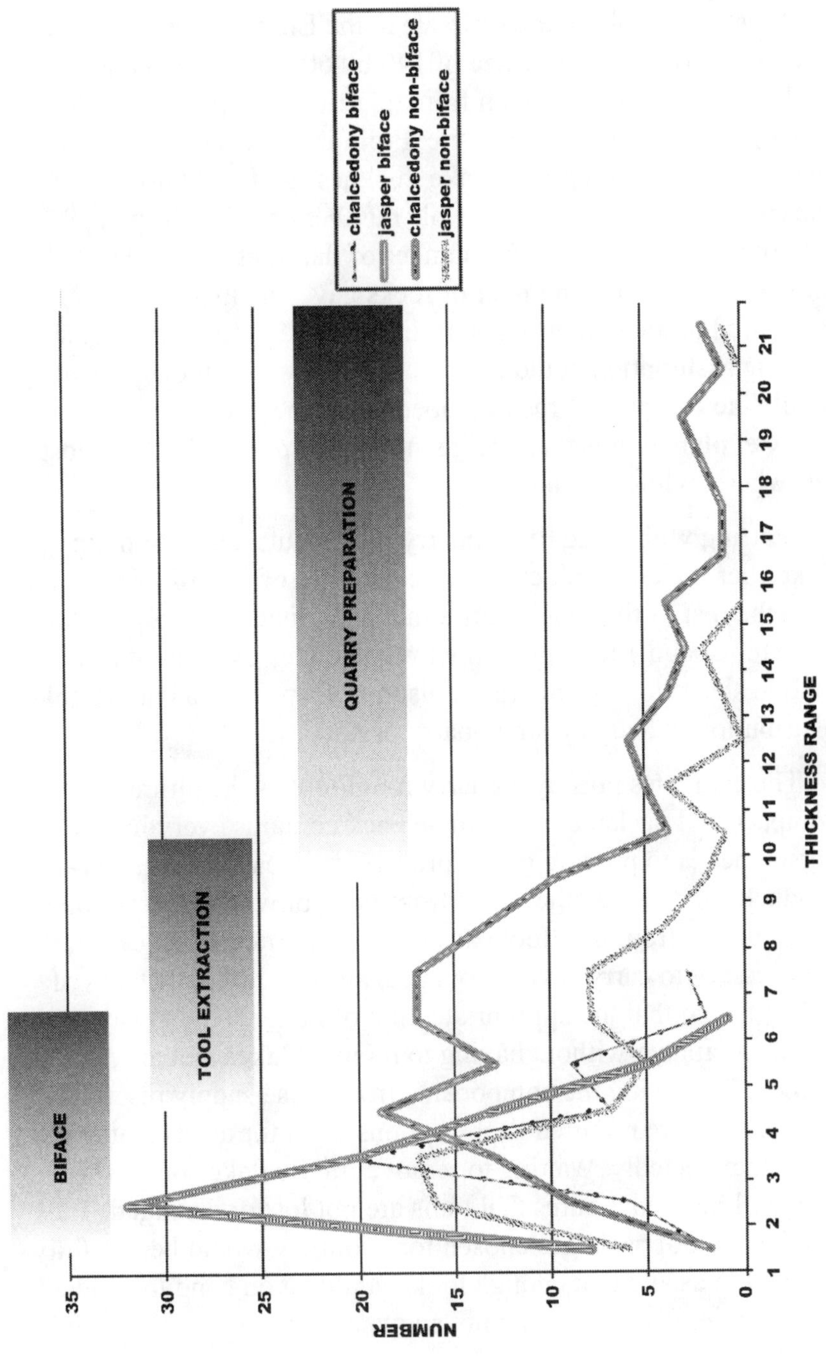

possible, but also accounted for the stage of reduction represented by the debitage.

I based my volumetric estimate at the Lindenmeier site on an average starting preform size of 140 by 60 by 45 mm, consistent with sizes of preforms taken from quarries, to represent the form in which stone arrived at the site. From the starting volume, I used the proportions of reported flake thickness and width to account for the extraction of the least possible number of flakes from a preform without waste. Dividing the number of flakes at the Lindenmeier site by the estimated number of rocks gave an answer of 54.7, compared to a volumetric determination of 51, although a new starting assumption could either narrow or widen the discrepancy. While the estimate of rock numbers could be in error, a more likely explanation is that a large number of projectiles were lost elsewhere, while hunting.

Starting with a bag of 30 quarry rough-outs, and assuming 50 flakes per rock, a production cycle emerges of 10 to 20 projectiles with the rest of the rough-outs either abandoned or broken, 100 flake tools, and 1400 unmodified waste flakes over about a half-inch in size. Stone from other Folsom sites reflects a comparable distribution, whether near a quarry or not.

The estimates portray a quarry rough-out as the ultimately compact tool package. That single pack contains everything a person needs to procure game, process the meat and other by-products, and carve other tools from bone or wood. Since tools are removed from the "tool package" in a particular sequence, it is necessary to carry several tool packages at different stages of reduction, so that the appropriate kind of flake would always be readily available without having to remove flakes that are not yet needed. Removing the components from these supply reserves only as they are required keeps options open for needs that may arise unexpectedly. Waiting to remove fragile flakes only when they will be used assures that tools are not lost or damaged. The number of tool packages chosen for a journey would be carefully planned to assure that enough tools would be on hand to conduct all the anticipated tasks, but not so many as to burden the group

Sectioning a rock

Volume	Portion	Thickness	Width	Length	Number
75,600	Quarry	45	60	140	
9,800	Preform	7	40	70	
711.4	63.4%	3.55			32.4
5176.2	31.2%	8.55			15.9
14639.0	3.8%	13.55			19.4
29749.9	1.1%	18.55			.56
3936.4	0.3%	21			.15
89642.9	**99.8%**				**51.0**

This mathematical division of a hypothetical preform relies on flake thickness data from the Lindenmeier site.

with excess weight. Surplus is not welcome if you have to carry it. Based on the Lindenmeier data, we can suggest that people's experience would have taught them how many tools would be reliably available from a pack load. While only one in three bifaces needed to become projectile points, it is quite possible for a biface to generate two or three points.

An analysis of the Fenn Clovis cache (Frison and Bradley 1999) in southwest Wyoming provides valuable insight into how readily tool flakes can be derived from bifaces. By comparing early stage preform thickness (16 and 17 mm) to finished projectile point thickness (6 to 9 mm), we see that a single series of 4 to 5 mm-thick overshots on each face could finish the preform. Folsom people continued the practice of making camp tools from large flakes taken from early stage bifaces. The subsistence strategy of maximizing tool flakes from a biface did not change when the projectile design changed. Tools made from edge-to-edge flakes are every bit as important as projectile points. Most modern knappers focus strictly on creating projectiles, and typically leave small, fractional flakes that would be unserviceable for dressing game.

Although edge-to-edge flaking technology on wide preforms invariably produces thin, flat flakes, few examples remain intact at sites. Flat flakes were commonly used-up by snapping edges off to create push planes for shaving bone or wood. Only the center portion typically remains. Thin Folsom flakes provide the functional equivalent of blades, but offer greater strategic flexibility. Snapped-off thin edges give similar working angles as burinated blade tips, without special set-up. Even projectile points were sometimes snapped to provide the special functionality of a freshly broken square edge.

Ultra-thin bifaces are increasingly recognized as a key component of fluting technologies. They rarely survive intact because they were literally used to death. Typically, only wedge-shaped fragments remain because any remnant of size had potential use as a reclaimed tool. Some unfluted Folsom points may have originated as a single ultra-thin knife. Because of their shape, ultra-thin knives were ideally suited for filleting meat for drying and transport. Later groups that adopted similar lifestyles used similar knives, probably for the same purpose. Some ultrathins were hafted, but many were used as hand-held knives. Evidence from sites near Lake Ilo, North Dakota suggests that a quarry blank destined for an ultra-thin knife was selected specifically for that purpose and not exploited as a tool core (William et al. 1997). In fact, there is more evidence of ultra-thin bifaces being completed near quarries than at other camp sites. Breaking an ultra-thin knife when replacement stone was not available could require making knives from bifaces intended for projectiles.

Clovis knappers made large platters that, at first comparison, seem to be related to ultra-thin knives. While the production techniques may have been similar, the use was probably different. Clovis bifaces are many times larger and heavier than ultra-thins—

Dis-assembled stone

Refitting helps us understand details about how rocks were taken apart. In this case, 17 flakes reconstruct the preform of a Clovis replica. Modeling this size preform predicts between 17 and 18 potential tool flakes of varying size.

in no way a tool to be easily wielded by one hand. The platters are most often found in cache contexts, and associated bifaces show strong evidence of being derived from deliberately broken platters. Both the platters and sections made from them are nearly always thicker than completed Clovis projectile tips. Thus platters take on significance as a form particularly suited for transporting stone in an intermediate stage of the reduction process. This stage is not seen in later manifestations of fluting tradition. At least one Clovis biface platter from a cache in Crook County, Wyoming retains an ochre shadow on the platter surface, revealing the use of a wide leather strap, probably used to protect it during transport (Tankersley 2002). Other cache bifaces from Clovis times show polished facets on high spots where the bifaces wore against one another. In part, the wrapping avoided breakage, but they may have been buffered to keep the pack from announcing the bearer's presence like a set of unwieldy cymbals.

Another Clovis cache, found by Dan Busse in western Kansas (Hofman 1995), shows that Clovis knappers sometimes harvested quantities of flakes too robust to have originated as bifacial flakes. Many flakes had been removed from tabular quarry stone by carefully orienting the impact to follow corner ridges. While they cannot be considered blades by the usual definition, they certainly qualify in the functional sense. By chasing prominent ridges, some blade-like flakes were removed from bifaces. These flakes are not overly thick, but are quite long and narrow. This cache is not only close to the primary stone source, it shows very little of the reworking usually associated with travel. Both large and small bifaces lack refinement, and camp tools are still unformed. While flakes could be derived later by reducing bifaces, a selection of convenient blanks picked at the quarry was significant. Fortunately, the cache had been removed from active service before the flakes were transformed from their starting morphology. The fact that uncompleted tool flakes were included in the cache shows that some tool requirements were best satisfied at the quarry. Exhausted tools from later cultures that are similarly robust are also likely to have been gathered at the quarry.

Disposition

As quarry-blank "tool-packs" are dismantled, their contents are left in the wake of the group's movement. Most excess bits stayed where they fell during tool manufacture, as pure waste. Many other flakes see a short life as expedient, unmodified flake tools before being abandoned. The relatively small number of formal tools, like scrapers, borers, or projectile points are transported, maintained, and reused for various spans of time. Eventually, they too are left behind through loss, breakage, or diminished functionality. Constant outflow from the production line forces the transition from process stage to process stage. A large kill site could substantially deplete the tool packs. Folsom butchering locations may have as few as one, or as many as 12 flake tools per carcass. When butchering was complete, people often took their tools to another location—extending the use of the tool.

While we don't know how long each tool was kept in service, there are some indirect means of estimating the relative span of service for each tool category. The same assumption that we used to detail a production cycle helps us to build a time-line, of sorts. It requires only the additional assumption that when a component of the tool pack was released from its "packaging," it represents another of the same tool exiting the cycle. It doesn't matter if a tool is lost, damaged, traded, or cached—that still defines the end of active use. If tools were not disposed of as replacements appeared, a surplus would be created. The only good reason for carrying a surplus is the anticipation of exchanging with another group of people at a scheduled rendezvous.

Whatever their purpose, caches are useful measures of episodes of the production sequence. Since caches are rarely found, we often think of them as having been hidden. Eskimo tradition provides a different model. Eskimo caches are stocked with carefully thought-out quantities of stores designed to assure the best chance of survival. When one must avail himself of the contents, he is obligated to replenish in kind. The location of these caches is necessarily public knowledge. The possibilities raised by relatively modern caches may be valuable clues that aid our interpretation.

Contents of Clovis caches range from nearly the full artifact suite, to just a few projectile points, or a small set of bifaces. The largest caches from any period of time consist of early-stage bifaces and can hold hundreds of items. At least some of the few known caches associated with fluting technology appear to be part of a burial offering, perhaps symbolic of a requirement in the spiritual plane. Other caches have cores and tool flakes, but few projectiles, indicating that they were ordinary camp provisions.

The Eskimo example suggests that caches may represent excess resources, placed where they would be most welcome in an emergency. There is little way to determine whether such caches represent fully-stocked or depleted conditions. Seldom are day-to-day tools found in caches, and they are nearly always located apart from camp or kill sites. Finished, unused projectile points are rarely found as part of provision caches, but tool flakes are seldom seen either. Cached Clovis bifaces incapable of yielding more tool flakes interrupt the normal transformation of bifaces into projectiles. One reason for not making projectiles is that enough were in hand already, but they could still be set aside for a future need.

In the course of a season, nomadic bands undoubtedly met one another and exchanged goods and service. The nature of trade tends to leave the resource needs of each party balanced. Since we assume the tool pack anticipates needs, trade would probably be limited to filling small imbalances rather than supplying another group with something they lack. Steve Holen of the Denver Museum of Nature and Science has studied Clovis use of stone in the Great Plains (2001), and sees regular transportation between 500 and 600 km. So little material moves beyond that range that he views exchange as being responsible for any greater movement of stone. Highly aesthetic, distinctive stones are most apt to travel the farthest. Distinctive stone from particular sources may have provided a useful way of communicating where a person had traveled or traded.

Many large Clovis platters were subsequently split and converted into projectile preforms, but the number of unfinished and unused projectile points suggests a surplus of weapon points relative to other stone tool requirements. Thus, some Clovis caches may represent disposal of surplus in a form that would provide a backup in times of need. Because the ecology was shared by many animal species, Clovis people could not be sure what animals they might encounter next. This uncertainty meant that a surplus of tools could have been inevitable from time to time because the quantity of weapons and tools needed to deal with a mammoth were likely different from those suitable for an antelope. Rather than carry a surplus, the excess might have been cached in a place that could be readily relocated. While relatively few caches have been found, they typically are far from a source of stone.

My proposed production sequence was modeled on the lithic inventory of a base camp because it best approximates a full array of activities over a all seasons. It is possible that resource use varied by season and focus of activity. For this reason, quarry locations, kill sites, and family campsites should be reviewed against the base camp model. Other issues involve distance from lithic resources or seasonal variations in game availability. Either circumstance could easily alter the starting pack load and the rate at which the stone would be used. Such comparisons are currently difficult because too few estimates have been made. Stronger models for estimating the balance of tools, population, and span of use will be needed to refine the estimates.

Many experiments have been conducted to assess the degree of utility inherent in each category of tool. Obviously, by retouching a scraper when it becomes dull from use, a greater tool life is obtained. Similarly, each separate tool category has potential for repair or conversion to other use. Repair of broken projectile points might actually double the number of points in service, depending on how the break occurred and how the portions accommodated rework. We can only guess how many animals a single projectile would bring down, but it surely would average more than one per point, unless many points were not recovered and returned to use. Since most found projectiles are retouched, we should assume that reuse was common. Even after the projectile was no longer usable for killing, it was often converted to other uses such as scraping or engraving. Even flakes that have no obvious modification were likely to have been used for a variety of tasks before finally being abandoned. Clearly, the makers of fluted points had an extremely well tuned, efficient system of use that gave them a great range of mobility despite considerable distance between opportunities for replenishing their resources.

J. Morrow (1995) noted that Clovis people in the St. Louis, Missouri area left progressively smaller tool flakes the farther they traveled from the quarry. This fits the concept that bifaces are thinned by increments over time. Since thickness of flakes that can be removed from a preform is limited by the preform thickness, successive flakes should be thinner. Unfortunately, the value of recording flake thickness has not been recognized by enough archaeologists. Recording data from unmodified flakes, which often are ignored, helps us understand the relationship between the quarry blank and its constituent parts. If established groups used stone so frugally, we can speculate that the first people to occupy an area may have been even more sparing in their use of stone since they did not know where or when the next quarry might be encountered. We might consider relatively complete utilization of stone as "inconspicuous consumption." When so little evidence is left at a site, the technological process is especially difficult to decipher.

Experiments in repairing tools damaged through use and attrition establish reasonable scales for the length of a production cycle. By cross comparing between the various tool categories, a normal range of tool life should emerge. Rare tool types, like ultra-thin knives, appear to have longer spans of use than those derived from the standard biface tool pack. Their rarity stems from being used so completely that little survives. Unfinished knife bifaces seem to be rare apart from quarries, since manufacturing techniques for thin knives contributed relatively few tool flakes.

The role of food

Stone tools were tied closely with the acquisition and processing of animal resources. If we can link food and processing requirements, a per capita lithic budget can be bracketed between high and low estimates. To my knowledge, this is the first attempt to quantify how long it might have taken to use up a given quantity of rock. The importance of the projection is mainly to help us understand the order of magnitude, not to settle on an absolute figure.

The book "The Paleolithic Prescription" by Eaton, et al. (1988) provides a basic idea of dietary requirements. At one end of the spectrum, Arctic Eskimos derived 90 to 95 percent of their calories from meat. On the other extreme, Australian aborigines only obtained 10 to 20 percent of their food as meat from small animals. Since buffalo are large and plentiful, and vegetation was readily available, we will make a rough estimation that 60 percent of calories are satisfied from protein and 40 percent from vegetation.

Bison, the major dietary component of later fluted point users, has about 25 grams of protein and 3.8 grams of fat per 100-gram portion (Eaton 1978:108). A fully-grown bison bull weighs between 1,400 and 2,220 pounds, while a cow weighs between 750 and 1,100 pounds. A typical 1,100 pound bison cow produces between 225 and 348 pounds of fresh meat (Frison 1978: 328). Drying the meat into jerky lowers the weight by a factor of ten, and a pound of pemmican is nutritionally equivalent to 5 pounds of fresh meat. From this data, we can project that an average grown bison can sustain a family of three persons for a month.

By reviewing kill sites, we can get a sense of how many points left service in proportion to the number of animals processed. This is not the same as looking at how many points it takes to kill each animal—that is irrelevant when salvaged points are merely pressed back into use. Considering the ratio of abandoned projectile points to the number of bison remains from eight sites (see page 181) shows a low of 0.33 at the Agate Basin site and a high of 2.1 at the Stewart's Cattle Guard site (Jodry 1999). The Cattle Guard site is probably the most representative because both kill and camp components are represented. On the other hand, inevitable loss

Nutritional values per 100 grams

Wild game:

energy	133.1 Kcal
protein	21.9 g
fat	34.3 g
cholesterol	67 mg

Wild plant food:

dietary fiber	12.6 g
calcium	1310 mg
vitamin C	30.6 mg
iron	6.4 mg
sodium	26.9 mg
potassium	424.6 mg

Dietary choises are not always voluntary, but they dictate the effort needed to maintain health of the population (Eaton, et al 1988; 78).

of camp tools in blow sand may have raised the proportion of abandoned items beyond normal levels. Projecting a conservative figure of four bison per person per year, and the need to replace 2.1 projectiles for each bison procured, leads to a per capita quota of 8.5 projectiles per year. Unrecorded losses of projectiles occurred when injured animals escaped the scene of the hunt.

Anthropological accounts indicate that substantial overkill may have been common. When going after a dangerous herd of animals, the accounting of killed animals is done after the dust settles. Raising the estimate to 10 points required per year still only demands 30 preforms per person, per year—an amount that easily fits our hypothetical grocery bag. The fact that most recovered points show retouch speaks to habitual reuse of points, as does the low rate of points per bison at archaeological sites. Since Folsom people did not stay long in one place, or often make repeated use of a site, scattering 60 rocks per capita per year over the landscape would hardly register as a significant presence. No wonder it is to difficult to find Folsom campsites.

Projectiles per bison

Site name	*MNI bison*	*MNI points*	*# points/bison*	*Site pose*	*end scrapers*
Folsom, NM	28	28	1.0	kill	--
Lipscomb, TX	56	30	0.54	kill	--
Cooper, OK	78	27	0.35	kill	--
Stewart's Cattle Guard, CO	49	104	2.1	kill/camp	98
Waugh, OK	5	2	0.4	kill	--
Lake Theo, TX	12	7	0.58	kill	--
Agate Basin, WY	9	3	0.33	kill/camp	42
Hanson, WY	--	24	--	kill/camp	136
total	***237***	***199***	***0.84***		

MNI = minimum number of an item

Camps can be identified by end scrapers, but the number of points per bison apparently is not related to camp activity.

Band size

Some indication of the number of people in a typical band can be inferred from the Stewart's Cattle Guard site (Jodry 1999). Excavated by Pegi Jodry of the Smithsonian, the site reveals discrete activity areas and shows that each family hearth had a similarly patterned distribution of tools. From the evidence, 20 to 50 people can be posited at a late fall kill of 50 bison. After the hunt, about 100 projectile points were repaired or replaced while hides and meat were processed. Stone at early stages of reduction from two geographically removed sources indicates that at least two groups worked cooperatively to kill a large number of bison in preparation for the winter.

Our previous estimate of nutritional value argues that the hunting event would have provided sufficient meat to feed the group for three months. Any isolated kills during the winter would undoubtedly have been welcome as a change of diet, but not critical to survival. As important as meat is the acquisition of hide for shelters and clothing, particularly moccasins, as well as bone for tools. Projecting a constant rate of two projectiles per bison leads to a yearly per-capita budget of eight projectile points. Since no other time of year is so critical to acquiring a good supply of meat as autumn, the estimate seems to be reasonable as a minimum yearly quota.

Load

If we look only at evidence from a site like Lindenmeier (see page 171), it might appear that one of every three quarried rocks became a projectile, and that the load imposed on the group would be three times the weight of an average rock for each projectile point required. However, the few known Folsom quarry sites reveal some behavioral patterns that may complicate our interpretation.

The Black Mountain Folsom site, near the Continental Divide in southern Colorado, suggests that a group of males (inferred since the usual camp tools were lacking) quarried stone and took it away with relatively little on-site activity. At the Antelope Spring site, though, the behavior appears to have been intensive reduction and replacement of partially exhausted stores. This contrast of extraction strategy is important to our interpretation of the larger pattern of ancient economy. Initial processing at the quarry should lower the mass of stone seen at campsites. Since both strategies are known, without more evidence we must assume that neither dominates.

However stone was gathered, it had to be transported to where it would be used and discarded. People would not have carried extra weight for fun, so there must have been some perceived value to each part of the pack load. If more than three quarry blanks were needed for every projectile used, the pack load would be highly biased to camp tools rather than projectiles. The fact that stones were not immediately converted to projectiles at the quarry clearly shows the value of what we so cavalierly call "waste flakes." In spite of our fascination with fluted points, it appears they were the *least* necessary part of the tool kit in terms of survival value. People were willing to carry bags of heavy stone in order to support the mundane activities of life, and the few grams added by a projectile point were an easy extravagance to tolerate.

Later paleo groups developed a different overall strategy. Cody people, near the end of the Paleoindian period, harvested numbers of tool flakes by dedicated reduction to edged and thinned bifaces near the quarry. A commensurate number of worn out scrapers were jettisoned at the same time. Appropriately, Cody peoples supplemented their tool stone needs by taking flakes from rocks picked up along their travels. As a result, more tools are made of local rocks, and cortex is increasingly common.

Flake tools

Since not everyone is thoroughly familiar with how stone flakes are utilized as tools, we need to consider what purposes have to be met. In estimating between 10 and 20 projectiles from 30 to 60 rocks per person per year, we have yet to consider the most important question of all: "What inventory of stone tools do bison hunters require?" Although there is no direct means of specifying how much use each tool would be put to, the flakes available to use as tools presumably average from four to eight per person per day.

Tools made on the most robust flakes generally receive the most strenuous use. Refitting of site debitage shows that even when tools are broken or abandoned for one task, they were frequently converted to another use before finally being left behind.

Just about any flake could have been used as a knife, but the evidence of such use may not be at all visible. My own experiments prove that a bison can be skinned by a single unmodified flake without producing detectable edge wear. Often, the location or context of a flake suggests that it was a knife rather than purely a discard. Other flakes that were deliberately modified to rejuvenate or trim an edge for better utility are easily identified. As a flake knife cycles from one use to another, each task tends to leave different evidence of wear, damage, or retouch.

Scrapers can look much like knives, but their use creates a different wear pattern. Sometimes the grit imbedded in an animal hide can wear and polish the edge of a scraper until it is well rounded and smooth. Depending on the task or stages of use, scrapers can range from delicate to robust. Many scrapers have steep, narrowed edges that show use in a socket handle, evidenced by contact wear and edge damage. Handles allow such aggressive use that the tip of a scraper can be snapped off by contact with the haft, something that would not be likely if the scraper were strictly hand-held. Delicate spurs often occur at scraper corners, perhaps due to shaping for hafting, but possibly to serve some other utilitarian function. Studies of Ethiopian cow-hide tanners, using obsidian scrapers, provide a high estimate of about 4 scrapers per hide (Jodry 1999:243). Archaeological data points to a ratio of 1.4 end scrapers per bison, partly because chert scrapers are more durable than obsidian, and partly because not all bison hides would be processed. A conservative estimate is from two to three end scrapers per hide.

Borers range from small protrusions from a flake edge to large "drills." The varieties that look like rods may have been used as fastener toggles, drills, or awls. Socketed hafts for projectile points would likely have needed heavy-duty boring implements like reamers. The nature of tools used for boring dictates that the tips

Tools made from flakes
(Examples from Antelope Spring site)

Gravers were made on thin flakes, including channel flakes.

Flake knives are often small and delicate.

Broad-bladed scrapers may not have been hafted.

What are usually called "thumb scrapers" are actually bits that were hafted in a handle. This example shows damage on the right side from such a handle.

Reamers are slender and relatively robust. This one has polish on the dorsal ridge and the edges.

are likely to be damaged beyond recognition at the end of their life span.

Gravers are often extremely delicate, needle-like protrusions, often made on another tool. Multiple points found on flakes may represent compass gravers for making designs. Another possible use of a graver is as a tattooing instrument. Inevitable damage, and the fact that gravers are subtle and difficult to recognize, obscures their original quantities.

Often, expedient conversions of core remnants and broken bifaces were used as wedges to split bone or wood. Pounding is usually evident on these tools. As in the case of other implements with robust use, identifiable remnants may be only a fraction of what were originally present.

Cleavers made of large stone are suitable for heavy dismemberment. They may have been destined to become other tools, but may have been just too cumbersome to carry, and easy to replace. Because of the destructive potential of the heavy use of cleavers, they tend to be made from unusually tough stone. Discarded cleavers of local stone show that they were not transported far and were not included in the long-term supply pack

Radial sections caused by deliberate impact at the center of flat flakes or tools are common in Folsom sites. Pie-shaped wedges with sharp edges and acute angles were evidently favored for a number of applications. Squared edges would have functioned well as small scrapers, planes, or as gravers. Use indications are usually too subtle to detect by eye.

Conclusions

Recognizing the life cycle of stone tools opens up myriad opportunities for deciphering patterns of use. An economy that supplies basic everyday needs remains in balance by extracting just the right tools at just the right time. Site remnants record the unique particulars of each individual camp, but also bear witness to remarkably stable trends over time and space that reveal culturally established subsistence patterns.

One way to visualize how the estimates illustrate daily life is to consider a band of 30 people—a typical estimate for mobile hunters. In an average day, one projectile would be pressed into service and the equivalent of one bison consumed. Various activities of skinning, butchering, processing, and tool-making would scatter 1500 flakes about the camp. It might be that each person made their own flake, in which case virtually every band member was trained as a flintknapper. The speculation seems reasonable, since the person doing the task knew the requirement best and could obtain the tool without interfering with other band members. To assure that projectile needs were met, particularly favorable preforms would likely be reserved for use by the most competent knappers. Other preforms would be available for practice by aspiring young hunters. Academic estimates of fluting "success" may be fruitless since the real goal was to acquire the skills necessary to supply the camp with stone tools. This model of lifelong exposure to knapping skills goes far in explaining the extraordinary competence exhibited by practitioners of fluting.

Maintaining such stable relationship between flakes, tools, and projectiles cannot come from haphazard dismantling of rocks. The next chapter reveals how people were able to extract just the right flake for the particular purpose at hand.

Chapter 8

The Language of Flakes

Flakes are ubiquitous, often comprising well more than 90% of the lithic evidence from an archaeological site. Yet, often enough, they are ignored in the analysis of a site because we have not learned their vocabulary. Instead, we tend to focus on the small percentage of artifacts that clearly speak to us of patterning and modification that is specific to a particular culture. Why should we assume flakes could not speak to us as well?

All fracture, whether man-made or natural, responds to the same laws of physics. But, while natures hand is haphazard, human decisions organize flakes in ways that are unlikely to occur naturally. To make the right flake on demand requires conscious planning, based on tradition and practice. Every aspect of a flake, from selection of the stone, to platform construction, to the knapping tool that removed it from the core, tells something about the people who left it behind. Archaeological evidence from many sites allows us to see how each flake is important to the reduction process. The following discussion is not restricted solely to fluting technologies; it may be useful for deciphering information from any lithic assemblages.

Flake characteristics

We will first consider how a worker visualized each flake and its contribution, regardless of the stage of work. Several descriptive aspects of flakes offer clues to what a worker intended.

Flake stiffness

The varied shapes of flakes make them difficult to sort, but boxing their major dimensions describes them most simply. Flakes proportioned like bricks are relatively robust. Flakes that are wide and thin like a sheet of glass, are more apt to break. Going one step farther, flakes that fill their boxed dimensions are stiffer than those that do not.

Flakes from the earliest phase of reduction are often boxy, with squared edges and flat faces. At the other end of the spectrum, biface refinement produces relatively platy flakes with acute edges. The reconstruction exercise in chapter 7 was successful because the work habits of Folsom craftsmen were so consistent that their flakes averaged 0.256 of the boxed volume. Setting a priority on stiff flakes most suitable for camp tools leads a knapper to continually seek out the most angular ridges; blade cores are the logical consequence of such a strategy. Making bifaces necessarily requires flakes to become progressively flatter and more acutely edged. If the worker does not commit to the proper priority at the right time, he risks failure.

Platform

For edge-to-edge flakes, the impact offset defines how thick the flake can be. The degree of offset is helpful for placing platform reduction flakes in the proper stage because it shows how big the preform was.

At any stage of work, platforms must be strong enough to take blows without shattering. From the strength of the platform, conclusions can be drawn about the type and size of hammer, and the stage of reduction. When the remainder of a flake is missing, the platform shows whether it was made to accommodate the heavy blows of an early stage, or the finesse of refinement. As bifaces are reduced, platforms generally progress from natural, to faceted, to reduced-by-battering, to reduced-by-beveling and finally, isolated.

Initiation

The depth of fracture initiation affects flake morphology. A shallow flake following a ridge will be long and slender. A deeper flake in the same location may expand greatly if the ridge is not very high. The width of a flake at the core edge depends on the concentration or dispersal of load at impact, and the distance of the impact from the core face. Hard-hammer impact starts a small, circular crack that then flares until the edge of the platform is reached. Soft hammers deliver force evenly enough that the flake detaches along a simple arc. Highly isolated platforms or greatly reduced margins provide the greatest chance of a flake being narrow at the edge.

Low angles of blow exaggerate the bulb of force because the flake tries to keep forces in balance by swelling. Especially high angles of blow have the opposite effect because they separate the flake from its core by bending, which creates a lip.

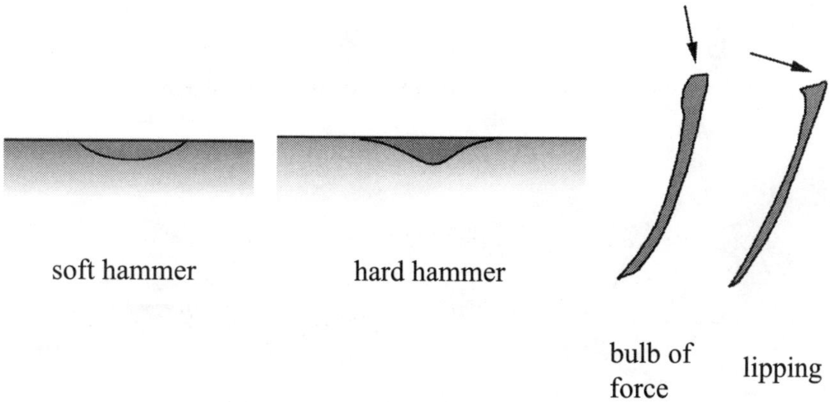

Hammer hardness shapes the detachement from the core, and blow angle shapes the first few millimeters of flake travel.

Flake thickness

Flake thickness is a good indicator of work progress because it is limited to about 40 percent of the parent core thickness. Thick flakes are physically impossible at late stages of work because the core is so much thinner.

After handling lots of flakes from workshop sites, it becomes clear that edges of bifacial flakes tend to align with the center plane of the flake. That fact allows us to estimate flake thickness associated with a flake scar by doubling the measured depth of the scar. Large flake scars from Clovis platter bifaces can be shown by this analysis to have generated flakes perfectly suited for tools. On the other hand, flake scars from most advanced bifaces do not indicate flakes thick enough for tools. Flakes removed in the concavity of flake scars are necessarily thinner than normal biface flakes, and may have most mass below the edge plane than above it.

Ridge chasing typical of advanced biface reduction, where the main concern is to make the overall surface even, reduces surface isolation that supplements flake thickness. Also, flaking into a concavity does not allow thicker flakes unless platform offset is increased, which tends to sacrifice biface width.

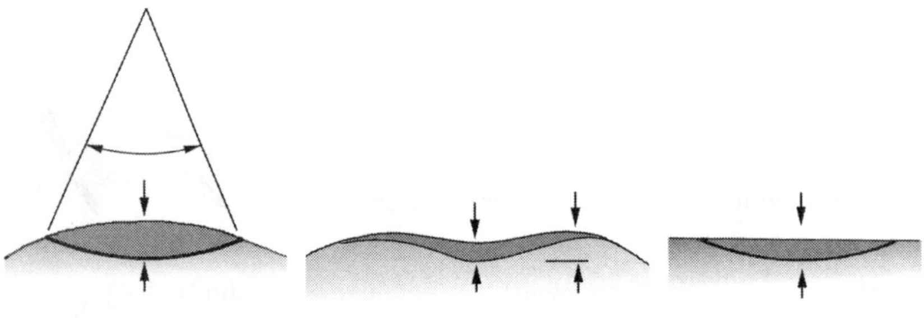

Estimating flake depth depends on knowledge of the initial surface shape.

Concavity

As the crack advances, the characteristic side-to-side arc at initiation flattens progressively until it is nearly flat at termination. A perfectly flat core surface causes a flake shape that is always the same, regardless of size, and is distinctly bulbous. The trough of the flake is limited to a few millimeters of depth, and the radius of flake scar concavity provides enough information to determine the missing surface.

Since the strongest scar concavity occurs at flake initiation, and the least concavity at termination, the change is usually distributed evenly along the length of the flake. Blades, because they keep a uniform cross-section, retain nearly the same concavity for their entire length. Thicker portions of flakes tend to be flattest because that is where they are most stiff, and greatest curvature occurs at acute edges, where resistance to bending is least.

Shape

When a flake travels deep enough, the core sides, rather than the surface, may define the shape. Otherwise, the plan shape is rounded. When flakes trend perpendicular to the biface centerline, they can be fairly long, narrow, and thin. Erratic orientation of flakes leads to asymmetric shapes with "arms" formed when the flake expands to follow ridges. Sequential flaking can create an asymmetry pointing in the direction of flake advance, especially when flakes are closely spaced. If the asymmetry usually points clockwise, it can be due to an individual preference or a cultural one. Consistent bias to one side could indicate handedness of the knapper, but the same result can occur if a worker advances from one flake to the next in the same direction. Overshot flakes often flare widely at termination, especially when the far edge is uniform and thick.

We learned earlier that flake shape depends primarily on stiffness due to the surface geometry above the fracture plane. Completely erratic flake shapes are relatively uncommon. More often, you can tell if the knapper was chasing ridges to refine the surface, and whether the flake sequence was selective or serial.

Overshot flakes were usually taken away as tool blanks, but scar patterns on the parent stone and occasional reject flakes testify to their presence or absence. The less variation in flake plan, the less range of process steps is represented at a site. Since each group of people had their own ideas about how flakes should be organized, their work should naturally be more or less distinctive. If nothing else, we can determine something of a group's knapping strategy by the characteristic flake shapes left behind.

Travel

The forward progress of a flake path is usually more complicated that the lateral expansion. High angles of blow cause a distinct bulb of force at initiation that may turn into a deeply rippled surface if the forward and outward forces are imbalanced. Perfectly balanced forces leave nearly imperceptible bulbing near initiation. When blow angles are too high, the flake path usually ends with an upward sweep, often accompanied by rippling. Flakes that encounter extra surface relief are driven to a downward arc of termination. The downward turn causes the flake to flatten its lateral curvature. Since fluting technology was geared to making nearly edge-to-edge flakes, scars typically reach well past the biface center, particularly where ridges channel the flake path.

Patterning

Flakes taken earliest in the process have no evident surface scar patterning, but each subsequent stage tends to show a greater degree of scar organization.

Process stages

Flakes must satisfy the requirements of each process stage before the knapper can advance to the next goal. Whether dealing with squared edges, or bowed surfaces, or twisted preforms—each requirement demands a particular kind of flake. By paying close attention to what is needed for each step of reduction, we gain crucial clues from the morphology of a flake. Some preforms can be moved through the stages in a gracefully uniform progression of refinement. Others may advance by awkward increments, whether because of intractable stone or less-than-skillful knapping. Overall, each stage leaves debitage that is progressively more uniform. When working with a large quantity of flakes, it is helpful to assign each one to a process stage. Physically laying out the flakes in columns representing process stages, with largest flakes at the top of each column, graphically illustrates the disassembly process. Whenever possible, it is helpful to compare the evidence gleaned from flakes against the negative remnants of flake scars on the parent stone.

Quarry blanks

When the parent rock was considered too big for transport, it was sectioned into easily carried slabs suitable for later modification into bifaces. By observing the largest flakes remaining, we can infer that large flakes, no thinner than 10 mm with roughly parallel faces, were required. Quarry blanks typically start with cortex or weathered surfaces, and minimal modification. The knapper does not want to carry parts that he obviously cannot use and so will strike off cortex, ridges, humps, and inclusions, both to make sure the stone is suitable and to rid the blank of obvious defects, like cracks, that would invite disaster on the trail. Heavy-duty scrapers often retain cortex. The first flakes from freshly quarried stone do

Spalls left at a site are usually either flawed or too small for use.

not have patterned flake scars on the surface, and platforms lack formal preparation.

Core preparation

Stone at the quarry often has very acute arrises, or ridges. If a biface is to be produced, each step of refinement will retain progressively flatter arrises. A core intended to provide nothing but tools may be irregular because the core could be continually re-oriented to get the best flakes. Regularizing the blank helps set it up for transport. It is also crucial to size the stone so that a suitable number of tool flakes can be extracted before it is exhausted. The upper surface of flakes from this stage show scars from earlier removals. Platforms for blows that take off remaining ridges and humps are generally unstructured. Rocks with large, unpatterned flake scars can be interpreted as cores. The ventral surface of flake tools can sometimes be recognized as matching the dorsal face of typical cores.

Core refinement

Few complete flakes are found at this stage because they would have been removed and used for tools. Flakes that are not removed from the site generally contain defects that the craftsman did not want to be bothered with. Flake surfaces retain some patterning, usually collateral.

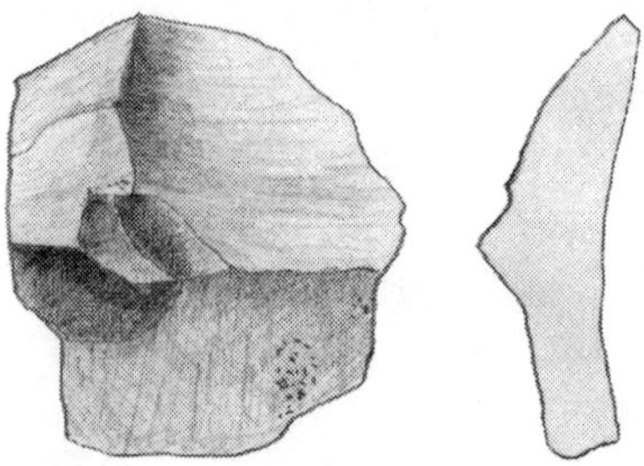

This early stage flake shows how the fracture removes a longitudinal ridge by deeply penetrating from edge-to-edge.

Early preform refinement

Although the transformation from core to biface seems indistinct, when a preform's thickness reaches that needed for bifacial tools, flakes removed from it are too thin to be very useful as tools and many are left where they were created. Flakes are then directed to refining shape and surface contour. Flakes at this stage are normally flat, widely spaced, and retain end-laps because they cross the centerline.

Now, flakes generally follow the existing contour of the biface, and average about an inch wide. Highly arced, thin flakes with negligible overlap have traversed far across the preform, sometimes reaching the far edge. Platforms are carefully placed and structured, causing flakes to have fairly standard shapes. Once the target thickness is reached, refining flakes that lower remaining arrises seldom cross the centerline and often fall far short. Flake

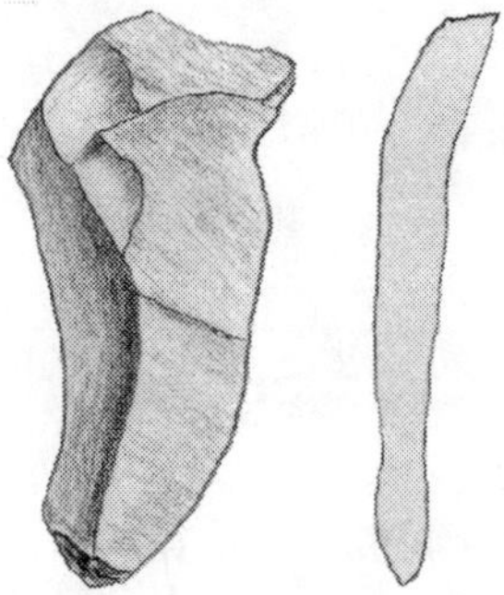

Ridge removal typical of early preform refinement leaves flakes that are symmetrical and reveal distinct arrises.

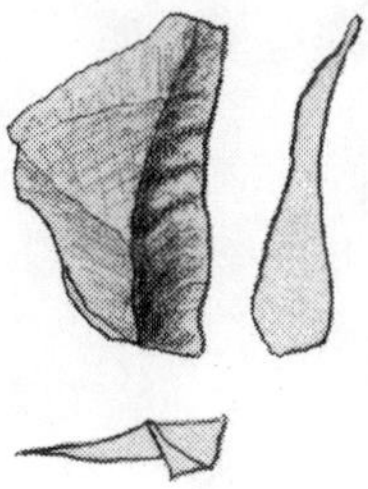

Overhangs from early stage preform flakes are part of margin preparation.

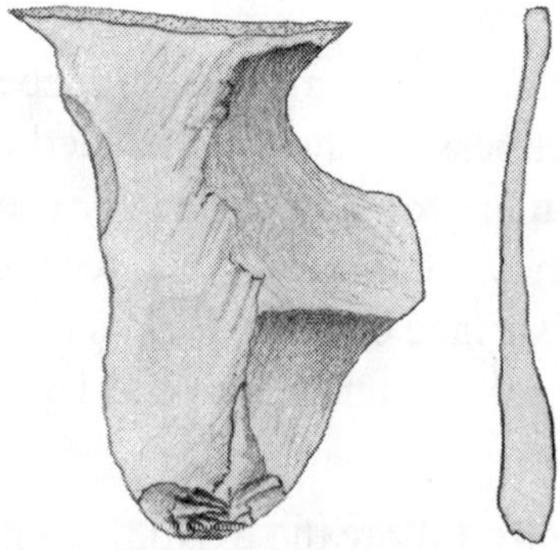

The tendancy of flakes to follow ridges causes this erratic shape on an overshot flake at an early stage of preform shaping.

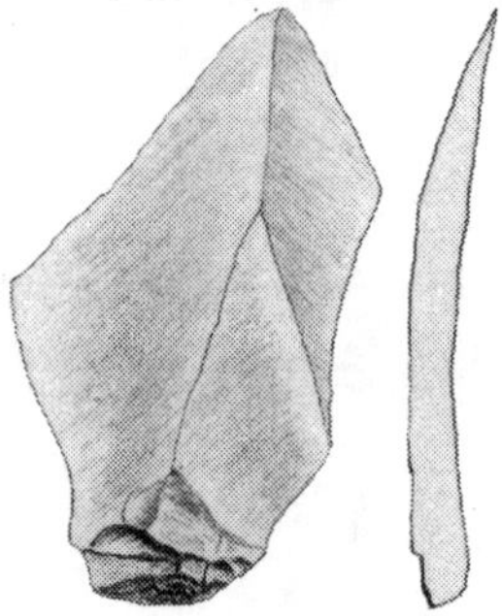

Early stage preform refinement shows distinct arrises, and orientation nearly perpendicular to the preform edge.

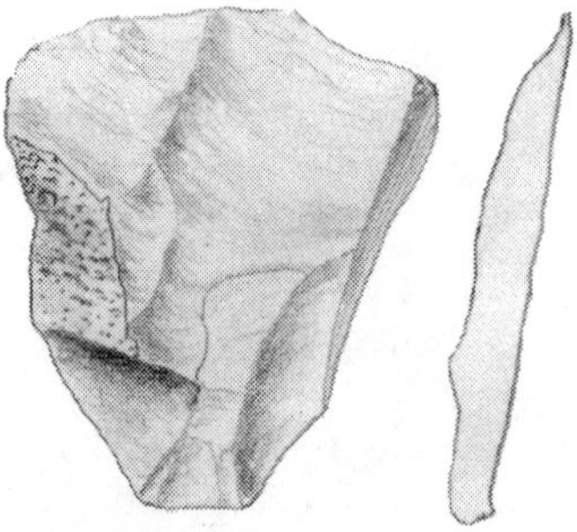

Uniform surface leaves typical rounded flake shape, even when the flake is thick.

ends are often rounded and scar origins on the remaining biface are usually obliterated by a later stage.

Late preform refinement

These acutely-edged flakes are simply too small and thin for tool use, and usually retain overlap scars on the surface. Platforms are carefully structured by pressure and are ground. Arrises are weak,

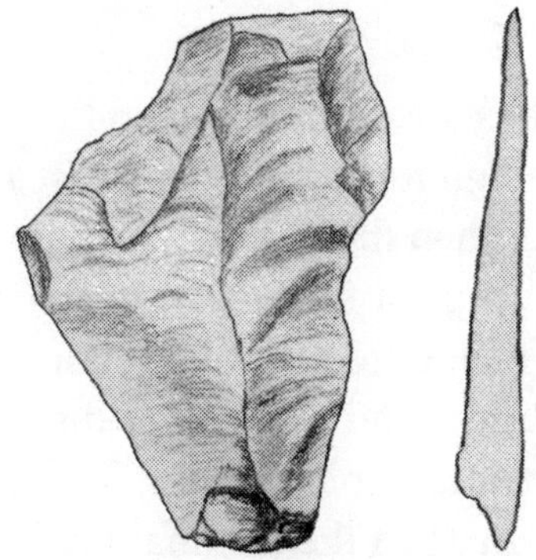

Ridge removal during preform refinement causes typical flaring flake shape and highly acute edges, because the surface is relatively flat.

but may be followed to develop surface continuity on the biface. Patterning is most common when surfaces are being regularized.

Edging

The edge of a biface is commonly made even and straight by selective removal of offending arrises. Flakes that give the biface a sharp edge are generally short, and usually have acute ends. There is seldom anything that makes edging flakes recognizable other than their small size and relative abundance.

Pressure flaking:

Pressure flakes are seldom more than a quarter-inch across, and are often destroyed by crushing against the flaking support. Platforms are proportionally small, and sequential flakes tend to follow arrises of previous flakes. Pressure flakes tend to have more ripples than percussion flakes.

Flake functions

Not all flakes can be clearly identified as having been made at a particular process stage. However, a flake's appearance often suggests the knapper's intent. Matching that intent against the apparent disassembly process helps in determining where a flake belongs. Because each tradition has a different recipe for taking rocks apart, we need to be careful not to mix rules.

Platform preparation flakes

Platform preparation flakes are common from the early stages, but may be difficult to assign to a particular phase of reduction. They are typically shorter than they are wide and very often were made directly underneath a prior flake scar, which gives them a distinctive wing-shaped feel. Often, a flake retains the remnant of an overhang from a large scar and sometimes displays traces of battering. If the offset is measurable, multiplying that number by 2.5 gives a minimum estimate of the original preform thickness.

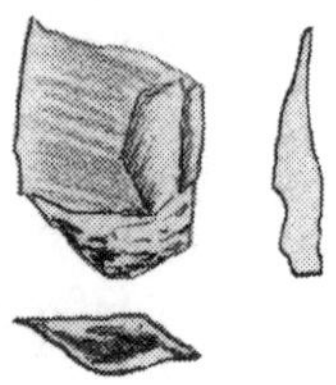

Flakes that move the impact location are small, but provide a way to estimate how much an edge was being moved.

Clipped corners

Squared edges are often made acute by clipping a corner ridge, either at right angles or at an angle to the edge. The resulting flake can capture the morphology and help us estimate the thickness of the quarry blank.

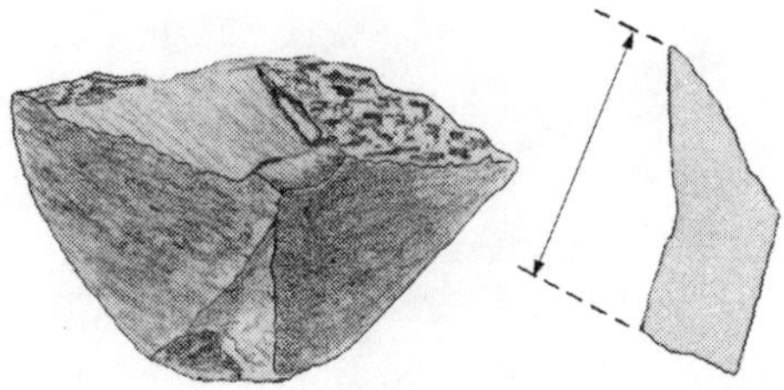

This flake was made to give the knapper a relatively acute edge to strike against

Ridge removal

Ridges can be serious obstacles for patterned flaking and are often removed by deep flakes. These are commonly transformed into scrapers or other heavy-duty tools.

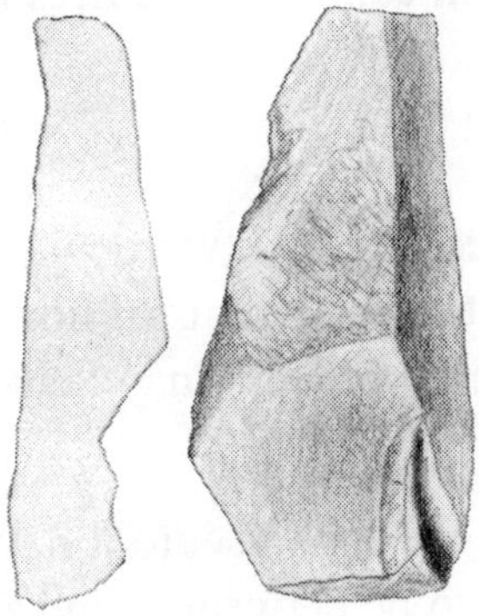

Flakes have tallest arrises when they come from quarry blanks. Note how the deep flake travel removes surface impediments.

Contouring

Long, wide flakes are useful for setting up the necessary surface control for a biface that is destined for fluting. While those flakes may be thin, they often have such regular geometry that they become the basis for more delicate camp tools, like scrapers and gravers.

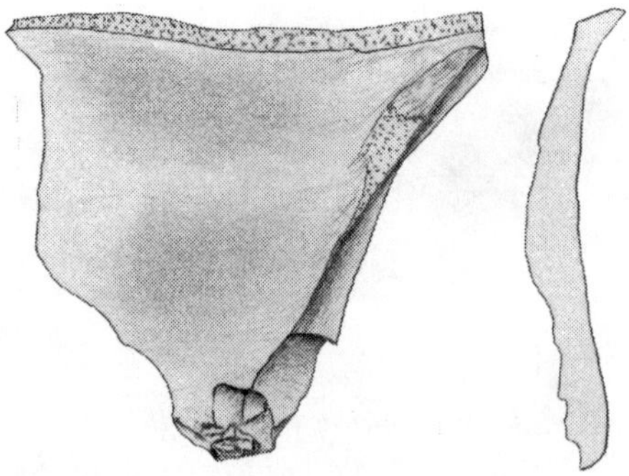

This contouring flake has an asymmetric cross-section on an overshot flake resulting from a slight clockwise advance from previous flake.

Damage

Accidental flakes are seldom patterned and often show shatter. Potlids and crazing from camp or wild fires typically obscure surface contours and retouch patterning.

Basic strategies

Many strategies for breaking rock were widely shared for large spans of time, but subtle details of how techniques were incorporated in the overall process may help in sorting lithic traditions.

In order to stand up to heavy use, tool flakes from most Paleo-Indian cultures needed to be between 6 and 8 mm thick. While flake width varies according to surface isolation, edge-to-edge flakes of 7 mm thickness tend to be about 50 mm wide when taken from bifaces. Spacing bifacial flakes near the normal flake width allows us to approach a theoretical maximum utilization of stone. If the flakes are all oriented at right angles to the long axis of the biface, then multiple sequences of flakes can be removed while maintaining a relatively uniform surface. Selective flake orientation, such as chasing ridges, drives the knapper to erratic patterns of non-standard flake shapes, and often leads to twisted or curved bifaces with uneven surface morphology.

Sorting flakes by similar form indicates what core or preform surfaces were like, even when the core is absent. Early in the process of reduction, flakes are relatively bulky and irregular. As cores are exhausted, they yield flakes with more and more regular shapes. While this may sound like common sense, the principle is only useful when applied systematically to a large assemblage of flakes, and compared to flake scars at various stages of reduction. Once flakes have been segregated by shape, it is apparent that the only significant difference between them is scale. We see that strategies for flake removal can be the same regardless of core size. The only significant restriction is that a flake needs to be large enough to function as a tool. Flakes that remain unretouched at the workshop location effectively mark the boundary between big enough and too small.

Careful ordering of flakes on a biface or a core serves to control surface development. Placing one flake alongside the other develops a facet that can be completed in a single pass of ordered flakes. Sequenced flakes naturally create aesthetic flake patterning by their relatively uniform size and spacing. That appearance evidently was valued in early times, as it is by collectors today. However, if staying with uniform sequencing led to an uneven biface surface, an early craftsman would give surface control a higher priority than appearance, at the expense of aesthetics. Understanding those kinds of priority rules is not easy. Analysis of the Fenn Clovis biface cache (Frison and Bradley 1999) shows that flaking patterns were often achieved by interrupting flake sequences to address problems that were developing at other parts of the biface. When the sequence was resumed, spacing was adjusted to fit in with the intervening scars in a way that requires detailed attention to flake order to recognize. Generally, the rule was to sequence from tip to base.

The most robust flakes found archaeologically are those that remove ridges and surface lumps. While they regularize the core surface, they also have characteristics that suit them well for durable camp tools. Cortex is common on large tools because the earliest flakes made use of the strongest ridges on the starting core. In fact, chasing ridges is a central part of the strategy for preliminary reduction. As long as the core has adequate thickness, it pays to organize ridges to facilitate more thick flakes. Spacing flakes at the maximum flake scar width automatically creates the strongest feasible ridges on bifaces. You can see the result on large Clovis bifaces that were cached because they were at the end of service as tool flake cores and ready for conversion to projectile points.

In regions where stone was available either in large or plentiful packages, chasing ridges led to another end. Bulky cores accommodate taking a succession of slender flakes from the perimeter. Each long flake (blade) scar leaves a prominent ridge to guide a neighboring long, thick flake. As the sequence progresses, the flakes become more regular and proportionally blade-like. Occasionally, the core top must be peeled off to allow more flakes to be removed. The distinctive platform truncation is called a core tablet. Blade cores are shaped by selecting the best ridges to make usable tools. Depending on stone availability, cores may sometimes be conical or wedge shaped. Because areas that have stone sufficient to support blade technology are relatively restricted, there is confusion over whether blades are useful in defining Clovis technology.

When adjusted for scale, cores and blades made by other cultures appear identical to those of Clovis time because the same relative conditions for fracture exist. Clovis blade flakes are normally thick and long, but discarded cores possessed enough mass to make plenty of smaller blades. If separate cultures produced different sized flakes, despite having identical morphology, the flakes must have served different purposes. It is possible that long, robust blades provided the use-life necessary for processing extra large animals.

Edge-to-edge flakes are essential for creating usable tool blanks. Anything shorter sacrifices length that is crucial for long tool life. An added benefit of edge-to-edge flakes is that they allow thinning a biface without the need to first edge the entire perimeter. It is possible to make a biface by using only feathered flakes, but critical tools requiring uniform thickness from end-to-end would

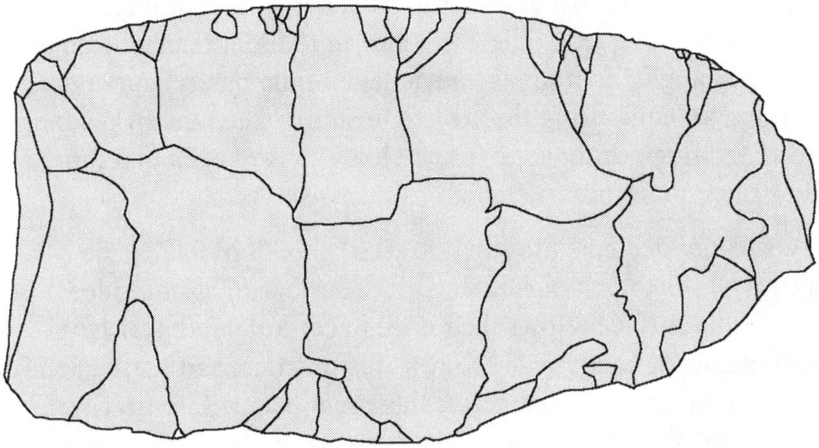

A

Biface A comes from a cache near Ft. Collins, Colorado, and is actually 30.3 cm long. Because of similarity to other large Clovis bifaces, it is likely that this is a core at the end of its useful life for supplying tool flakes, However, after being split, it could easily be transformed into two sizeable projectile points.

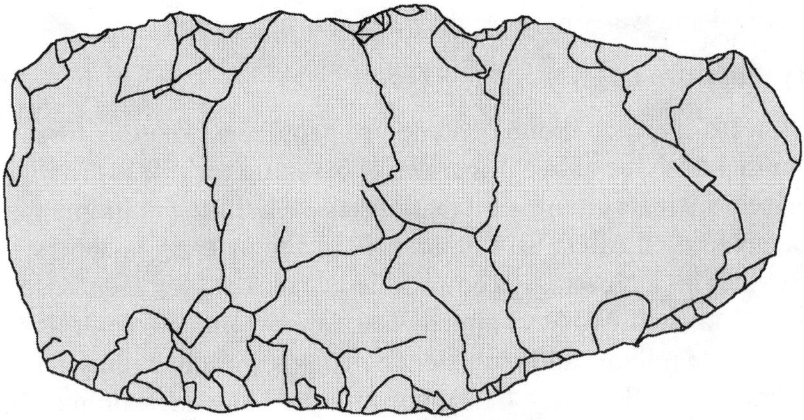

B

Biface B is a replica only 12.1 cm long, but scaled to the same image size as Biface A. The similarity between the two occurs because the same strategies were used for each.

be lacking. When bifaces are large, it is best to orient flakes at right angles to the long axis. The last series of tool flakes can be directed at a slight angle, both to take advantage of the natural curvature of flakes for contouring the projectile, and to keep the flakes long. Flakes from opposing edges are inclined toward each other to allow them to terminate smoothly.

Once a biface gets too small to yield good tool flakes, the strategy shifts to contouring the biface surface. Chasing ridges to contour the surface allows the biface to control strategy. Many modern knappers obviously follow this practice, and consequently have trouble creating distinctive flake scar patterns. Artifacts that exhibit strongly patterned flake scars tell an analyst that the craftsman was in conscious control of flake placement, spacing, and orientation. Closer spacing of flakes, and thin, feathered flakes that strip away ridges, predominate for the final phases of projectile production. Percussion flake scars transition toward the rounded plan shape typical of flat surfaces, and pressure techniques are used to smooth the surface between major scars. If thickness has already been established, scars may not even reach the center of the preform. Flakes from this final stage are most likely to be left at the site of manufacture and are most in danger of being removed by wind and water because they are so insubstantial.

Tradition

Of necessity, each group of stone-age people was more or less familiar with the above principles of breaking rock. It is the way those principles are ordered that makes each lithic tradition unique. Deciphering the details of lithic activity among early knappers provides an increasingly complex description of how their business was conducted. We have already learned how process controls encourage reliable and consistent outcomes without calling on extraordinary dexterity. Craftsmen also tend to work within fairly narrow zones of possibility, favoring what is comfortable. However, group standards of production were flexible enough to allow individuals to make implements different from another. The way tasks or work stages are sequenced reflects a subtle mix of process control and social convention. For example, the point at which tool flakes are removed, within the life cycle of a rock, varies through time as the natural environment changes the requirements for tools.

Flake scars from bifaces associated with fluting traditions indicate that flakes were highly organized both in orientation and spacing. The diagrammed relationship between core thickness and flake thickness suggests that relatively few tool flakes can be wrung from a biface less than 15-mm thick without sacrificing potential for a projectile tip. The frequency at which that dimension is found in caches of Clovis bifaces indicates those bifaces were past the stage at which they could supply camp tools.

Variation in core and flake thickness by tradition

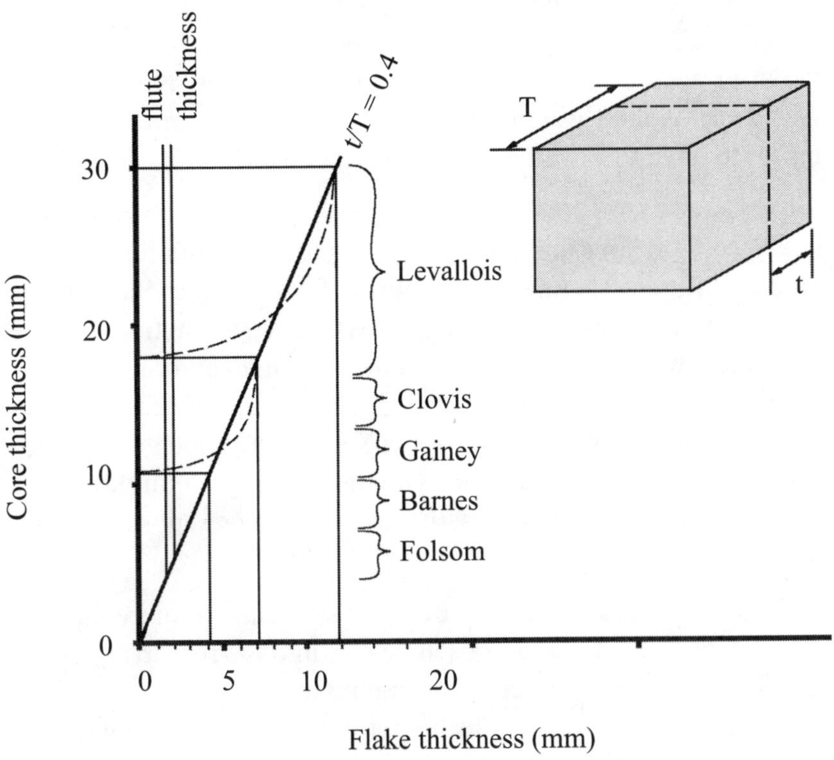

Lithic traditions can be sorted by their preform thickness which, in turn, influences typical debitage thickness.

Artifacts in various stages of manufacture from the Fenn cache in SW Wyoming (Frison and Bradley 1999) illustrate how at least one group of Clovis people organized their work to use flakes differently at progressive stages in the life cycle of a stone. The conversion from platter to biface preform occurred when overshots no longer would provide sufficient thickness from that stage on. Finished Clovis points are typically 6 to 9 mm thick, and have very closely-spaced flake scars. It is no accident that flake scars are increasingly close as the projectile approaches its final form. The decreased flake width reflects a corresponding decline of interest in flakes as tools and an increasing dependence on flakes to sculpt shape. It appears that the Clovis process of manufacture was not necessarily linear, because successive operations could be carried out independently of one another. Inter-related priorities were constantly balanced, and decisions made on what to do next throughout the construction. Although it may be argued that the Fenn cache is extraordinary, it is still not clear that "ordinary" Clovis projectiles did not pass through the same degree of excellence on their way to discard. In contrast, Folsom workers appear to have followed a draconian schedule of work that left little room for improvisation.

Projectile shapes and flake patterning are most often relied upon to differentiate one lithic tradition from another. Subtleties of shape and scar patterns are unlikely to serve a function, so they may reflect personal preferences. Among 19th century makers of gunflints in Britain, workers reputedly could identify each other's products even though, to an outsider's eye, they appeared identical (Forrest 1983:96). A similar ability to distinguish an individual or band's handiwork might easily have existed among stone-age hunters.

Details of process controls are more subtle and require larger sets of data to evaluate, but reach deeper into the reasons for what people chose to do. Choices of knapping tools and sequences of work stages reflect on how the ancestral technology base was adjusted in response to new experiences. Inventing entirely new methods each time retooling became necessary would lead to much more variation than is found in the archaeological record. Even

Flake sequencing

Two sets of flakes scars on a point from the Fenn Clovis cache have the oldest scars shaded lightest to show how sets of sequential flakes can be orchestrated to produce an overall visual effect. Fluting was finished prior to the patterned flaking.

when, in Paleo-Indian times, final projectile forms seem to depart from their predecessors, the underlying processes can usually be demonstrated to be nearly the same. When the particular point of departure is identified, it should tell much about the reason for the variation. Cody projectile points possess distinctively faceted faces produced by pressure flaking against an inflexible chipping stage. Projectiles with parallel flake scars, immediately after that time, lack shouldered hafts and faceted faces, but otherwise retain the Cody technological signatures. The most evident adjustment was the removal of the rigid support during pressure flaking, which allowed flakes to arc from edge-to-edge.

Expanding typology

By considering the full scope of stone tool design, it is possible to distinguish work of one time period from that of another, or even between the work of contemporaneous people who lived different lifestyles. Traditional typology describes an artifact's geometry and its implied function. So few artifacts remain as originally conceived, that the tools are often cataloged differently in their discarded state. Although typology has long been useful, it has potential for even finer distinctions of knapping practice.

An artifact acquires its *form* partly because of a mental template of what is "right", but also because that form allows the tool to do its assigned task. A tool made initially in a traditional pattern can acquire the appearance of a tool of another culture by the time it is discarded, as the tool's function is maintained while damage is repaired. Consequently, it is common to see many forms repeated in different cultural assemblages.

The knapper's physical behavior of supporting work and manipulating tools is *technique*. Fracture trajectories and scar morphology provide reasonably unambiguous clues for identifying technique. A single attribute is not enough since it may be caused by more than one way of knapping. Yet, combining evidence from debitage, flake scars, and platforms allows us to describe how an individual in the distant past may have used his tools to create a desired form consistent with a particular tradition. Each flake fits within an overall strategic pattern of disassembly.

As technique creates form, so *strategy* guides technique. Flake scars and debitage reflect how rules and priorities guided knappers decisions. To make the right flake in the right order, there had to be some kind of procedural instruction. Each tradition had its own rules of order and spacing that an individual might follow more or less consistently, depending on how compelling the rules were. Therefore, it takes many examples to properly characterize a tradition's technique. Many artifact outlines are identical, but were produced by completely different combinations of technique and strategy.

Context is how various elements are combined in relation to each other, and that is based on culture. It is quite feasible for several cultures to share an artifact form or even a technology, but it is unlikely that they would share every trait. People that live as a coherent group go about their normal tasks in an agreed-upon way. That gives their products an identity, which can be distinguished from those of other people. Including technique and strategy in an expanded typology, shifts the emphasis from the discarded tool to work in progress. Even better cultural identifications are possible when a full complement of artifacts, and their relative frequency of occurrence, is available to portray their context at a site.

Conclusions

Anthropolithic forensics reveals that patterns within an archaeological assemblage persist because site occupants habitually used particular knapping tools with familiar techniques, and applied consistent strategies, all learned from their elders. Once those patterns can be correlated to the environment of the time, we have some basis for describing a tradition. Because flakes tend to be left behind more consistently than other categories of artifacts, they are among the most useful clues for deciphering those trends and extracting meaning from them.

A lithic technology arranges its work steps to support a specific lifestyle in a particular geographic area. Attending to the smallest level of detail increases the opportunities for understanding a wide range of behavior. Differences between processes exist because novel solutions were found to meet new requirements. As long as new needs were encountered fairly steadily, the technology could develop and mature. Technological discontinuities can be expected when environmental calamities demand drastic accommodation. Whatever the impetus, each accommodation broadened the ability of succeeding generations to react to shifting environments. Strategies that persisted did so because they were stable. Judging by the archaeological record, the more generalized strategies were most successful because they could adapt to changing conditions most freely. The disadvantages of highly structured processes required for fluting eventually led to abandonment of Folsom points.

Each tradition left artifacts uniquely suited to cope with their environment. Those distinctions may be so stark as to be unavoidable, or buried in the details, but they must exist. Differentiating lithic traditions would be futile if it were only to verify differences and place cultural labels on materials. The real payoff occurs when motivations for adaptation can be identified. Then, the archaeological story can move beyond the dry recitation of chronology to explore the most immediate concerns of ancient people.

Chapter 9

Lineage of Fluted Points

Through the course of this book, we have seen how to characterize *what* archaeological remnants of stone were used for, learned *how* they were produced, and have then drawn conclusions about *why* they were important. We have progressed from understanding individual flakes, to deciphering the special process recipes that constitute a tradition. Each step of interpretation rests on forensic evaluation of available facts. By now we know what fluting is and have some ideas about how it was accomplished, but the motive remains unclear. Attempts to determine a lineage bring up new questions that are even harder to resolve than those we have tackled so far.

Hardly any reliable dates have been collected for the majority of potentially ancient sites. To remedy this situation, we need to supplement a few key dates with some assumptions about how lithic technology may have matured through time. Perhaps the early lineage is more like a thicket than a tree because of multiple points of origin.

What purpose did fluting serve?

Basal thinning is required to accommodate hafting for virtually all projectiles. It would seem that fluting was an extraordinary refinement of basal thinning that gave extra advantages for hafting. Whenever we find bifaces being converted into Clovis points, we see that prominent basal thinning was started before the rest of the artifact was refined. Clovis points seem to have either relatively narrow bodies with trough-like flutes, or fairly flat bodies with plane facets that create a standard cross-section angle at the base. The difference in structure appears to indicate a different hafting systems. Faceted flutes provide plane surfaces on either face for the hafting to seat firmly. Scratches observed on obsidian Clovis points imply that mastic kept the projectile from shifting during use. Unfortunately, no evidence has yet been uncovered to show us what the actual hafting system looked like.

Why do something as risky as fluting?

We know from scars on large bifaces that overshot flakes were made routinely, and risk is low for something done habitually. The scarcity of those flakes in sites means that they were expended as tools. Because flakes were regularly made to be as long as possible, there was no obstacle to fluting projectile points by direct percussion. Other technologies that did not maintain edge-to-edge flaking usually considered end-on flaking as unsuitably risky.

Is every fluted point part of a fluting tradition?

Basal thinning does not necessarily come from the base, but lateral flakes make it nearly impossible to standardize a hafting taper. Flutes can be generally defined as longitudinal basal thinning flakes. However, to qualify as a fluting tradition, longitudinal basal thinning must be consistently made by the same method as the biface, following a deliberately isolated surface, and accommodating special hafting needs. A flute that solves a one-time tapering problem is simply an anomalous characteristic.

The primary styles whose inclusion in a fluting tradition may be questionable include Mesa points from Alaska, South American fish-tailed points, and Dalton points. While some of these points have flutes, the flutes occur too infrequently to be a defining attribute. For example, only about 10 percent of Mesa points have flutes. Mesa points have such thick bodies that flutes are occasionally created when a thinning flake clearly follows an isolated surface, but that ridge was not made for the purpose of fluting. While the scar looks like a Clovis flute, it is not a routine design requirement. However, it raises the consideration that fluting may have emerged when heavy-bodied projectile points needed tapered bases for hafting. South American fish-tailed points are usually too thin for a flute to follow an isolated surface, but anecdotal accounts indicate that some specimens were nearly fully-fluted in a way that could only be accomplished through deliberate preform structuring. The flutes on Dalton points are generally short enough to argue that surface isolation was not important.

Both Mesa points and South American fish-tail points were fluted with the same kinds of percussion tools used for bifacing. Dalton points, however, appear to have been fluted by pressure.

Following these arguments, South American fish-tailed points appear to be part of the fluting tradition, while Mesa and Dalton points are less likely related. Other instances of flute-like basal thinning should be similarly tested before including them within a fluting tradition.

Where did fluting traditions originate?

By looking at how humans first populated remote areas of the world, we can get an idea of how colonization of the Americas might have progressed. As Jared Diamond has pointed out in his book *Guns, Germs, and Steel* (1999), even in Australia and the Pacific islands, where significant oceanic barriers were present, people managed to arrive nearly as quickly as if there were no obstacles. As soon as they set foot on new land, they inevitably left a visible imprint by altering stone, plant life, and the landscape itself. Without pre-existing populations in the way, humans can expand at a precipitous rate. In fact, the Clovis-first model is based on rapid colonization of a continent with no prior occupants. But were Clovis people the first, and where did they come from?

Along the Beringian margins, several potential precursor technologies can be identified, but none of them has a clear advantage in explaining Clovis origins. Most Arctic stoneworking traditions use microblades extensively, while the Clovis camp sites discovered thus far have been largely devoid of such forms. The few northern traditions, like Mesa, that contain apparent quantities of bifaces, share hardly any special traits with Clovis technology, except that virtually all Arctic peoples used their tools to exhaustion when they were distanced from fresh rock. Outcrops of good quality flint are common on the North Slope of the Brooks Range in Alaska (Kunz, 2003:39-40), but appear to be scarce elsewhere. Economical use of stone is not a technological trait as much as a behavioral strategy. Because tool stone could only be procured during the few warm months, it needed to be conserved throughout the rest of the year. By the time spring arrived, not enough stone remained to give much clue as to how tools looked at the onset of winter. Of necessity, miserly practices were the hallmark of most early stone tool traditions in the far north.

A given chunk of rock can be dismantled by either of two strategies utilizing edge-to-edge flakes with very different qualities. A microblade is a very specific type of end-to-end flake that sections the rock into slender blades. Since the remaining exhausted core cannot be used as a projectile point, a projectile must be constructed by compositing the blades onto a slotted bone point. Edge-to-edge flakes taken from a biface are somewhat irregular, and more readily used as camp tools than for composite tools. However, the remaining core can be used as a projectile tip after those flake tools are made. The microblade option pervades the coastal regions, while biface flakes are present inland. Immigrants from Asia possessed micro-blade technology as early as 30,000 years ago, but lacked edge-to-edge biface flaking. It seems unlikely that a bifacial technology could have traversed the gauntlet of pervasive microblade use without incorporating them.

While coastal settlement models have been proposed for both coasts of the Americas, sea craft and marine adaptation can only be logically postulated on the basis of later cultures. Evidence of such early occupation on 60 to 90 extra miles of now submerged continental shelf is simply out of reach by our present means of data collection. Coastal models imply a reliance on marine animals. Harpoons have tended to have flat bodies, whatever time period they were used, and that may have encouraged the use of edge-to-edge flakes.

Immigrants coming from Europe would have needed to follow seasonal ice sheets stretching from Spain to Newfoundland at the end of Solutrian times, nearly 15,000 years ago. They possessed every technical attribute needed for fluting technology, but relied on blades as a primary basis for tools. The scarcity of good quality stone in large nodules along most of the eastern coastline may have caused a shift away from blades in favor of edge-to-edge flakes from bifaces. Their technological background would have prepared them to alter their strategy without having to change their techniques.

If we presume the Americas were populated before fluting technology appeared, an earlier people would have proven their ability to cope with local environments. Indeed, relatively few Clovis points have been found from the West Coast and the Basin and Range area west of the Rocky Mountains, and stemmed points appear at very early times. A few sites along the East Coast of America are dated much earlier than Clovis. For these possibilities to be properly evaluated, we need to date and define them better.

What should a pre-fluting technology look like?

Cultural Virus Theory, developed by Ben Cullen (2000), points out that objects cannot replicate and evolve by themselves; it is ideas or knowledge that must evolve. Artifact assemblages allow us to identify which actions were most highly controlled during knapping. Tight controls are associated with low variation, while loose controls necessarily lead to more variation. Controls were key to achieving the desired outcome, and are assumed to reflect technological solutions from an earlier time. We do not know how long it would take to invent a process control, or precisely how the knowledge would be incorporated into a cultural tradition. However, increased layers of dependant actions, and complexity of necessary process steps, suggest that development must have taken generations before projectiles clearly recognizable as "Clovis" were more than random occurrences. By tabulating some of the controls used at early, mid, and late stages of tool production (see page 220), we can suggest a progression of technological elements that may be compared against potential precursors to fluting traditions. Additionally, some key attributes of those traditions provide clues to the potential degree of relatedness.

The founding technology for fluting traditions should include the ability to make edge-to-edge flakes by baton percussion. Edge-to-edge flakes are possible only with knowledge of proper support at the opposing edge of the core, control of flake thickness by impact offset, and use of surface isolation to avoid wavering in the flake path.

Bradley and Stanford (2004) have recently proposed that points found along the length of the eastern seaboard may have origins in the Solutrian period. An early variant of Suwannee points found at the Page-Ladson site in Florida have been tentatively dated at nearly 14,345 RCYBP. Page-Ladson points have triangular shapes similar to points from the Meadowcroft site in Pennsylvania, and points dated to over 17,000 years ago at the Cactus Hill site in Virginia. They also resemble points recently reported by Guadalupe Sanchez from northeastern Mexico (2004). Caches of oversized Simpson points found in Florida are reminiscent of Clovis shapes, although they lack flutes, and provide another possible branching of technology aside from Clovis.

All of the Eastern candidates for pre-fluting have relatively flat, thin bodies as a consequence of baton percussion. Mesa points from Alaska have thick bodies, but even the earliest dates do not precede Clovis. The morphing from flat bodies to thicker ones may tell us something of the earliest hunting practices. Relatively recent Arctic marine hunting used harpoons with relatively wide, flat weapon tips. Some of the earliest stone points along both east and west coasts have exactly that characteristic, and are often found in proximity to large Pleistocene sea mammals like walrus and seals. Moving inland may have encouraged heavier bodied points to deal with longer haired and tougher skinned prey. Basal thinning of these points could logically lead to fluting.

Why has it been so difficult to spot a Clovis predecessor?

In considering the origins of American lithic technology, we can presume that the critical evidence exists, but we haven't been able to either find or recognize it. Many claims have been made for pre-Clovis occupation in various parts of North America, but no tradition of fluting exists outside the Continent. If, as it appears, the Clovis tradition was invented here, someone must have already been here to invent it. Although claims for pre-Clovis traditions are problematic, they must be considered at least as likely as having a new technology spring magically out of nowhere. But how could people have migrated here so much earlier than Clovis and not left recognizable evidence?

A number of sites in Nebraska and Kansas, dating from as early as 18,000 years ago, suggest that the use of bone for making tools may be partly responsible for the scarcity of evidence (Holen, 2003). Mammoth skeletons would have offered a plentiful source of durable bone tools that may have substituted in part or whole for stone. Bones at these sites are found broken in ways not likely to occur naturally. Impact cavities are consistent with a baseball-sized hammer, and are widely spaced along the length of a long-bone. No predators besides man have been known to modify bone in that way. Two-inch thick bones are thoroughly broken alongside delicate ribs that remain intact. Successive blows were placed and organized in just the patterns we are familiar with when stone cores are broken into useful sections. Sections of bones might have been taken away to be modified into tools that had little chance of surviving millennia. If the sites represent scavenging of bone rather than processing a kill, there may have been no reason to leave stone implements behind—assuming stone tools existed at all.

Because we know so little about the earliest lithic technologies, they could easily be masked by their resemblance to artifacts from

Process controls

Full-length fluted points
(Barnes, Cumberland, Folsom)

Stage	Controls
late	anvil support blunted/ground tip longitudinal surface isolation face at a time platform offset unwavering preform crest
mid	uniform thickness

early	full-length flakes
	  offset platforms
	  surface isolation

Clovis

Stage	Controls
late	fractional flutes pointed preform tip
mid	opposing edge support salvage from split flake cores
early	face flattening longitudinal surface flattening platter biface cores

Precursor controls

full-length flakes
  offset platform
  surface isolation
  isolated platform
opposing edge support
baton percussion

Attributes

Full-length fluted points

parallel fluted faces
expanding sides
ground haft area
delicate retouch
standard width

Clovis

wedge-type basal thinning
bowed sides
ground haft area
casual retouch
variable haft width
flutes
caches
some blades

Precursor forms

Simpson

flat body	overshots	bowed sides
caches	some flutes	ground haft
wedge base	short retouch	

Suwannee

flat body	overshots	wedge base
triangular	some flutes	

Solutrian

flat body	overshots	converging sides
caches	blades	

later periods. Linear progressions of development seem natural, but they reflect adaptation to an arbitrary succession of environments. The fact that very early artifacts from South America so closely resemble artifacts from Archaic periods of the desert West is because the living conditions were similar. Diversity of prey species prior to the end of the Pleistocene may have encouraged people to generalize their hunting strategies with tools that were indistinguishable from tools of later times. Pre-Clovis artifacts reported from Meadowcroft, Cactus Hill, and Page-Ladson illustrate how easily similar designs can be confused with those of another time.

Not only are artifacts sometimes difficult to recognize, many have been displaced or destroyed by nature. Very little of the late Pleistocene landscape remains. Valleys have been filled by sediment, and hilltops eroded away. For this reason, the odds of finding important evidence by accident are quite small. Evidence of activities along ocean margins is now deeply submerged. It is hard to justify exploring deposits of pre-Clovis age in which no artifacts have been discovered, but even negative results contribute to our understanding. Deep water exploration shows promise because the depth of submersion relates to the age of occupation. In several instances, digging deeper than the "oldest" layer of a known site has led to unexpected artifacts. For example, digging below the water table at the Gault Clovis site in Texas produced flakes that might precede Clovis technology (Collins, 2002). Since a lithic technology includes an entire array of tools, as well as projectile points. Lacking a distinctive projectile point, the rest of a pre-fluting technology might be difficult to recognize.

While knowing where to look is difficult of itself, there are reasons to suspect that population densities were too low to leave high concentrations of artifacts. Large predators like sabertooth cats and the short-faced bear would certainly have intruded on kills made by early humans, and would have considered humans prey as well. The extinction of browsers inevitably led to the demise of specialized predators because the number of predators is inextricably linked to the number of prey animals. This may be

Comparing point types

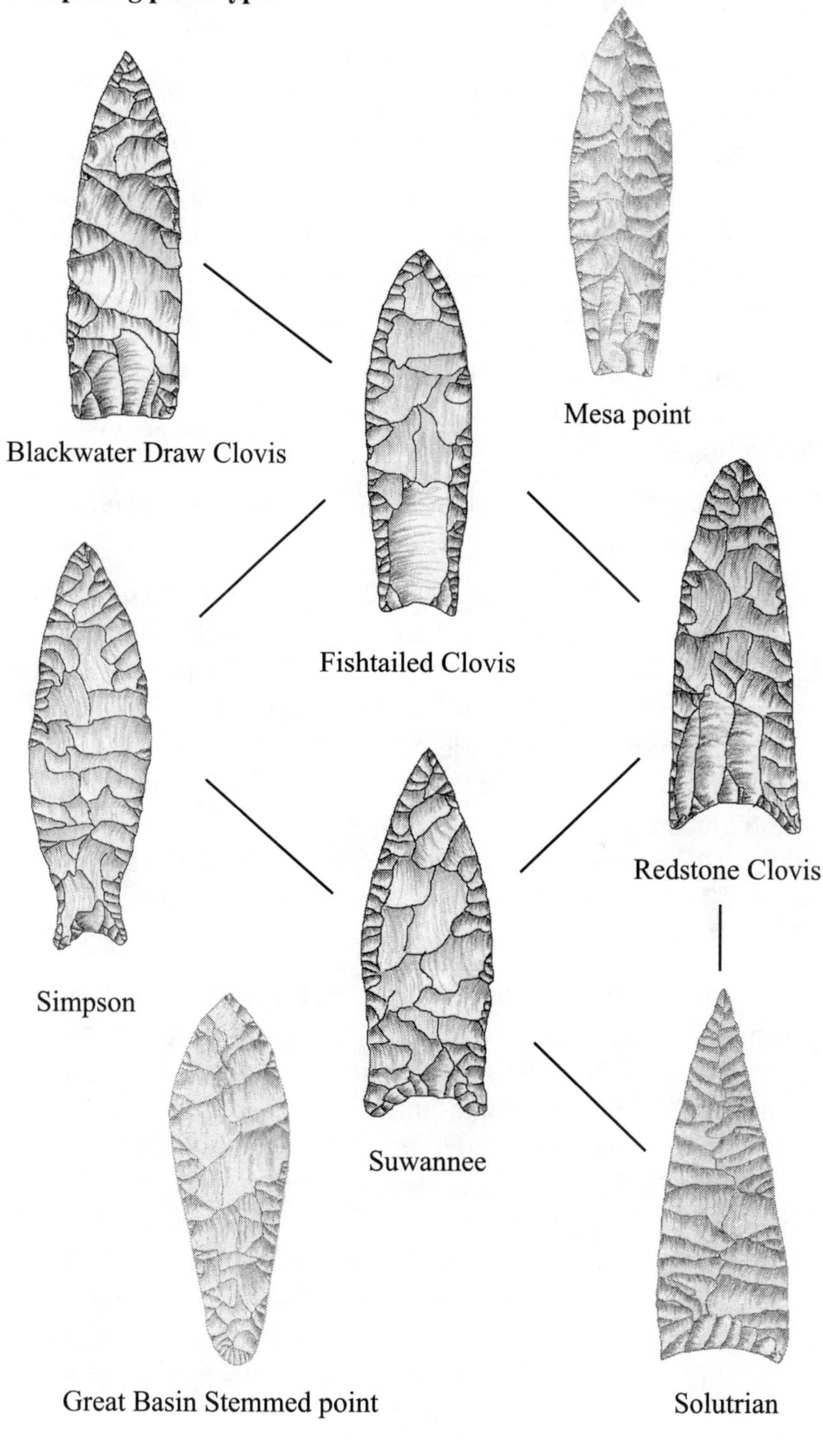

why human populations become visible once the large predators are gone.

Most of the initial research into earliest lithic technology in the Americas has been colored by the tacit assumption that a single group of people was responsible. Increasingly, we are learning that the diversity of artifacts, coupled with genetic studies, point to multiple migrations from many sources over thousands of years. Those possibilities may confuse us for a time, but they may make the story of pre-history much more interesting.

Until sites can be dated quickly, easily, and economically, claims for early occupation will be contested. At the same time, better dating methods will undoubtedly justify more claims.

What technological developments did Clovis require, and how long might they have taken?

The fact that Clovis people used fluted points at least 13,000 years ago is unassailable. But can such a complex technology spring into existence without some period of gestation? Depending on which dates one is willing to accept for the peopling of the Americas, it might have taken less than a hundred years or nearly 7,000. Virtually any other lithic technology can be seen to develop gradually from an earlier way of making tools. Clovis points exhibit a novel technology completely out of place relative to anything preceding it. If the evidence exists for a normal developmental phase, it has not been generally acknowledged.

Technological innovations appear to follow a logical and often necessary sequence. Each successive development depends on mastery of a more basic level. Although a few shortcuts may be advanced, most of the principles must be learned as extensions of a prior level of technology. Not only can we represent the progression as a series of levels, each level appears later in the archaeological record. Since the stages are logical advances and represent actual sequences through time, we can deduce that fluting technology must have evolved incrementally as improvements on culturally embedded norms. As long as the environment is stable, a successful technology tends to change very slowly. We can only estimate a minimum time of development by comparing against known transformations of knapping practice.

While the 5,000 years and 3,000 miles of ocean that separate Solutrian overshot technology from that of Clovis seem to be impossible barriers to overcome, the differences in making tools on either side of the Atlantic can be explained by generations of small, incremental changes. Developing Clovis technology immediately after people arrived in the Americas from Asia requires even greater change at an unprecedented rate, and with a baffling lack of developmental phases.

At the end of the Pleistocene, Northeastern Asian bifacial technologies, typified by unsophisticated bifaces made by hard hammer percussion, had been in place for tens of thousands of years. Living conditions and available game were not significantly different than existed in North America. Lacking the strong impetus for change by a radically different environment, incremental change should have occurred slowly enough to leave successive tool styles closely related to each other. Eventually, soft hammer percussion displaced the use of hard hammers without seriously altering the look of the technology.

The next logical improvement would have been to reduce the core margin through battering and beveling to create a marginal loading point that was thoroughly controlled. With this improvement in standardization, flakes of specific thickness would be available on demand. Not only is this necessary for simple camp tools, it creates reliable blanks for derivative forms.

With thickness under control via impact location, the next phase of improvement would be the quality of the flake surface. By attending to the lessons of biofeedback, people could match hammer and core stiffness to make relatively ripple-free flakes. Through experience, they could then manipulate preform support to control flake trajectory. Given enough time, flake removal becomes increasingly efficient, effective, and reliable as cultural guidelines tune-in to optimum conditions for flaking.

Just such a developmental sequence culminated in Europe nearly 20,000 years ago, when Solutrian knappers mastered overshot flaking. Solutrians were among the earliest to use this practice, but no other group used it as a mainstay until early Clovis times. In a developmental sense, overshot flakes reflect habitual use of edge-to-edge flakes to flatten bifaces. While it is tempting to see the function of such flakes as a requirement for forming elegant projectiles, the real benefit was in extending utility of a given volume of rock to increase territorial mobility. The fact that overshot techniques declined as populations settled into narrow ecological niches indicates that the underlying purpose of the technique was to supply tools during extended travel.

In the absence of full-length fluting, basal thinning is nearly always designed to create a consistent wedge cross-section. Introducing tip support not only lengthened flutes, it transformed the hafting paradigm to deal with parallel faces. When long flutes were eventually abandoned, wedge-style bases came back into favor. Clovis dates span only 300 to 800 years and later full-length fluting styles lasted about 500 years. At a minimum, it should have taken as long to develop Clovis technology as it took to transition to full-length fluting. If there were no impetus from a changing environment, it could take thousands of years for the technology to transform.

What makes fluting unique?

Stone sources in North America are scattered erratically, so to avoid being tethered to quarries people needed a way to make their stone last for an extended time. At the quarry, flakes can be made in any number of ways because the most suitable could be gathered and used. The farther people wanted to travel, the more critical it was to maximize the utility of each flake. As we learned earlier, fractional flakes, which do not reach the far edge of a biface, are tapered, and tend not to last long. On the other hand, flakes that reach edge-to-edge have relatively uniform thickness and start with enough length to provide longer tool life.

One option would be to carry selected flakes from the quarry, but that subjects long flakes and edges to possible damage.

Fractional flakes are fine for relatively local trips, but there are at least two other options for effectively dismantling a core that provide better economy and flexibility over longer periods away from a source of stone.

Microblades might look like a good, economical use of stone when one must wait out bad weather and can't afford to carry large quantities of stone, but as the climate allows mobility and better access to stone, the equation changes. The investment in building a composite projectile blade by carving bone rods to receive inset blades is high. Rejuvenating stone tips would be easier, as long as new stone is available. If the stone occurs in relatively large chunks at the source, large blades would be more apt to be made than small ones. Small rocks that could only yield small bifacial projectiles might better be turned into many microblades.

Removing very thin flakes nearly the full width of a biface produces the functional equivalent of blades. By relying on such maximal flakes, people could travel farther from a quarry with the same quantity of stone. Combining the advantages of bifacial technology with the advantages of blades maximizes the utility of a given quantity of stone. Very few traditions of stone working based their entire strategy of flake production using overshot flakes. The fluting traditions took an extra step by thinning their projectile points with longitudinal flakes. Of the later groups whose strategies approached those employed by fluting technologies, most were on the Great Plains where stone scarcity and migrating game provided the same impetus that the older traditions had responded to. It appears that fluting and edge-to-edge flaking are part of the same solution to an unusual need for using limited stone supplies for extended forays.

Clovis caches reveal that stone quarried from remote localities allowed people to routinely travel great distances, up to hundreds of kilometers during the life of a flake tool, much less the life of the toolmaker. The spread of fluting technology might be related to the relative freedom of movement that accompanies effective utilization of stone. Following the abandonment of fluting, fractional flakes from biface cores regained favor as people restricted their range of movement enough that they could readily replenish their stock of tool stone from nearby quarries.

Why is fluting so widespread, and varied in thickness?

As early people made their way to the Americas, their access to resources that could be used as tools was constrained by both geography and climate. In light of how cultures adapted such radically diverse technologies, it is amazing that fluting could be so uniformly distributed to every part of the continent. A miserly attitude toward stone could be explained by harsh Arctic conditions, if Clovis technology was invented on the way south from Beringia. However, previously established occupants could have adapted to climatically induced environmental shifts in exactly the same manner. Proposals that Solutrian technology arrived from Europe at an early time rely on just such a change.

The powerfully adaptive strategies of Clovis technology were suited to carry people through varieties of ecosystems. What I take to represent earliest Clovis technology is more basally thinned than dramatically fluted, and that is also the style most broadly distributed. It is also the scarcest of the fluted points, which suggests a rapid spread of technology. As later generations set about accommodating unique regional requirements, the technology transformed to be more in tune. Each of the transformations point to *adapted* occupations, where people made a commitment to live a lifestyle compatible with their newly adopted homeland. At the Gault site, for example, there seems to be little to distinguish Clovis inhabitants from later Archaic people, other than the weaponry. Regional patterns persist because people are constrained by their neighbors, even when they migrate to new areas. The wide variation in fluted point morphology may in fact reflect regional or environmental accommodations at local levels.

Based on unpublished evidence from the Mockingbird Gap site in New Mexico, Clovis adherents appear to be highly adaptive. Projectile points run the gamut from fingernail- to hand-sized. Presumably, any edible resource was fair game. Strategies of reclaiming and exhausting stone were suited to maximize their range and unfortunately contributed to the paucity of evidence available to archaeology.

West of the Rocky Mountains, stone sources are especially varied due to complex geologic structures. After the first wave of fluting technology, the hunting strategies so suitable to the East were not well matched to the broken geography of the West. Herd animals could not be depended on to be in predictable places at predictable times. Dispersed resources, combined with a multitude of isolated ecological basins, negated the advantage that edge-to-edge flaking offered in open regions to the east. At an early date, geographic constraints encouraged people to operate in relatively local universes, where resources could be utilized predictably. Only scattered occurrences of fluted points are known throughout the basin and range physiographic provinces. The move to stemmed and notched styles as early as about 11,400 RCYBP in Idaho probably reflects a corresponding flexibility of resources, and changed hunting strategies, based on the geography. As more people occupied the region, resource access depended on maintaining good relationships with whoever controlled a source, and people tended to make do with whatever stone that was within their range.

In areas west and south of the plains, where obsidian was available, Folsom people tried obsidian but seemed to dislike its fragility. Perhaps part of the problem was that it did not lend itself to durable camp tools. In the face of larger amounts of obsidian in Mesoamerica, however, a strategy of ultra-thin fish-tail shaped projectiles arose. This option grew naturally from ultra-thin knives and used basal thinning by fractional flakes. The lack of step terminations, in favor of feathered terminations, argues that anvils were not used to facilitate the fluting. When longitudinally-thinned preforms were subsequently narrowed to make a stem, the narrowed scars covered most of the remaining width, typical of most fluted points. Though other stone sources were encountered on the way south, the thin, fluted stem was retained until fluting vanished. Once again, the earliest fluted form is widely distributed and short-lived. Perhaps the lack of antler in the Southern Hemisphere made fluting technology impractical.

Gainey and Barnes styles were the thinnest style of fluting in the Midwest, probably because most stone in that region was not suitable for achieving Folsom or Cumberland standards. However, since stone sources were widely available, people were not tied to quarries and did not need to be stingy with their preforms. Work near quarries tended to move quickly toward finished forms, and edge-to-edge flaking was most important at early stages, while later stages made use of fractional flakes.

Fluted points approached the limits of achievable thinness primarily where sufficiently high quality stone allowed routine success. During Folsom times, there may have been little difference between cultures that used fluting. In fact, if high quality stone had been uniformly available, Folsom points might have been used in every region. Heat treating was not used to improve the success rate for thinner points because it causes flakes to break into sections when they are removed. Because the fluting tradition was founded on maximizing the length of intact flakes, the effects of heat treating would have been very detrimental. It also explains why obsidian was used so infrequently by fluting traditions.

Grasslands of the plains had relatively little stone of appropriate grade for fluting, but carefully planned hunting rounds allowed people to replenish their tool stock from scattered quarries, usually near the mountains. Stone selected at a quarry had to be used sparingly for hunters to span the distance to the next source and provide enough tools to harvest bison along the way. Edge-to-edge flaking was the most efficient way to extract appropriate sized flakes from cores and led naturally to Folsom extremes of thinness.

Cumberland points cluster around especially high quality stone sources along the west flanks of the Appalachian Mountains that effectively tethered the style to the quarries. In the Northeast, Crowfield and Holcombe points are cramped in a small area of distribution presumably limited by stone qualities that permit such thin dimension.

While fluted points varied in thinness according to stone quality, other variations in shape are widely distributed, and may reflect other cultural divisions that we can't explain for now.

Significant sources of stone

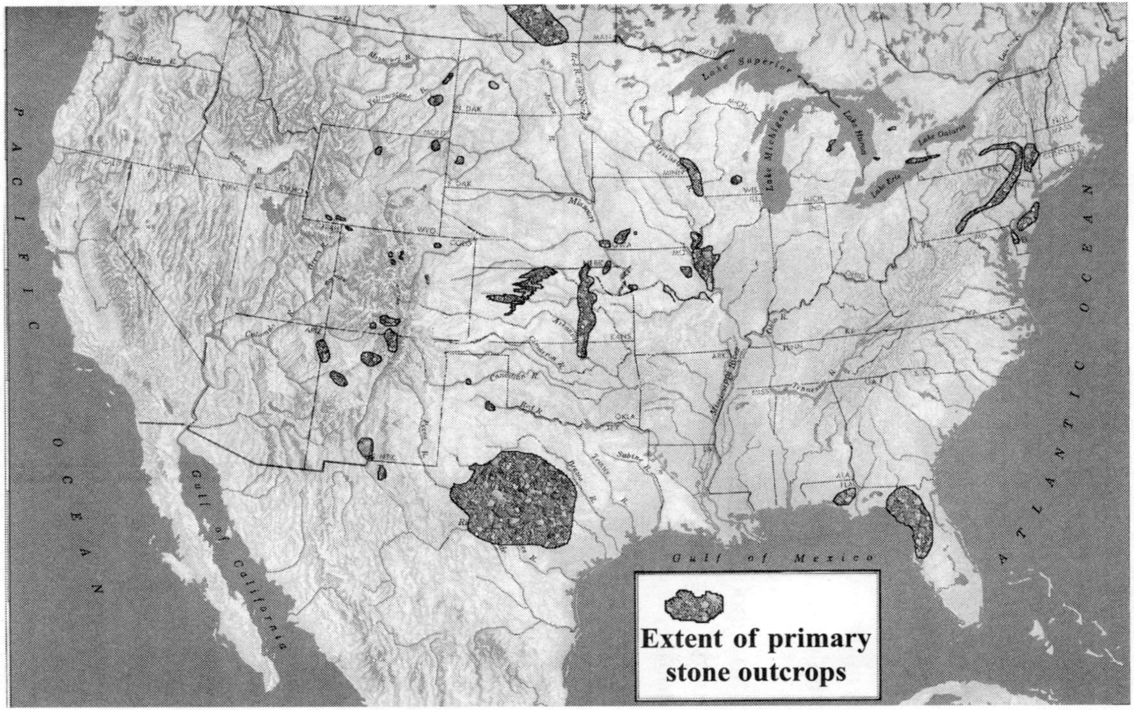

Stone sources used by fluting traditions are sparsely distributed and often are not readily available without considerable effort.

How did the end of the Pleistocene impact lithic technology?

The glaciation that preceded the end of the Pleistocene lasted longer than any other in human history, and brought unprecedented global changes when the ice melted. There can be no doubt that substantial changes in behavior were required to allow continued survival. As little as forty years may divide the end of the Pleistocene from the start of the Holocene. Half of the worldwide warming took place in a span of only fifteen years. The abruptness of climatic change would have emphasized the shrinking of viable ranges for many species of animals. Normally, incremental changes allow pioneer populations of plants and animals to replenish decimated territories. Without that option, many kinds of animals may have found themselves without sufficient threshold of territory to support genetic diversity. The impact, including similar extinction of large species, was felt all the way into South America. Technological traditions that had served well for thousands of years had to be revamped to accommodate a completely new environment.

Heightened seasonal variations caused by loss of the northern glacial barrier meant that plants were exposed to unfavorable growing conditions in the mid-latitudes. Arctic blasts would have killed many temperate species, and hot, dry periods would have killed many conifers. As plants died from the climate changes, new growth would have been restricted to relatively narrow ranges. Animals could readily move to better climates, but needed forage that was slow to appear in newer locales. Bison would have started to move in larger herds to follow shifting grazing pastures. Mastodon may have found it very difficult to transition from dying conifer forests to scrubby new forests that hadn't time to mature. Specialized grazers like mammoth, reindeer, horses, and camel would have been squeezed just as hard to find suitable food because of rapidly shifting ecozones.

The fact that Clovis expansion roughly corresponds with the extinction has been used to argue for Clovis cultures hunting megafauna out of existence. Yet, if over-hunting were responsible, we might expect a corresponding plentitude of hunting artifacts from this time period. In fact, remarkably few sites show a clear correlation between humans and even a fraction of the 34 species that went extinct. Whether humans were partially responsible for the extinctions or not, changes in climate, flora, and fauna were unavoidable. After the shift, many environments completely disappeared, leaving adapted animals nowhere to survive. Regardless of the reason, loss of megafauna was accompanied by necessary adjustments in technology. We have to look past assigning blame to recognize that the accommodation manifested itself as the fluting tradition. Edge-to-edge flaking is the mainstay of the new technology, while fluting is the identifying characteristic.

Loss of major prey animals not only impacted humans. As game became more scarce, people must have had tough competition from large predators. Film documentaries of the predator pecking order in Africa today give a sense of what life may have been like for small groups of early hunters when they downed fresh prey. In fact, the presence of big carnivores at the height of glacial impact may have kept human populations in check. A typical defense is to carry a large-bladed spear over one's shoulder so that if attacked, a weapon is at the ready. Jungle people of the Philippines are not hunters, but carry short spears nonetheless, to ward off tigers. Massai cattle herdsmen in Africa carry similar weapons to deal with lions. It is possible that large, cache-grade Clovis points were only needed for the period of time that people were under threat by large predators, and were dropped from the tool kit once those predators went extinct. Non-disposable defensive weaponry up to two-inches wide takes a completely different system of mounting than the more frequently encountered two or three-inch long points. In fact, the large styles have flat-faceted bases, while the bases of narrower styles are more trough-like.

Archeological interpretations of Clovis life are so heavily biased toward explaining their dependence on hunting mammoth that we can easily forget that the association is not proven. If we acknowledge the evidence of extremely high mobility for Clovis people, we should realize that, while they occasionally killed or scavenged mammoth, they would have been obligated to exploit whatever game they encountered to flourish as they did. Why concentrate on game that was both dangerous and too big to eat and process before it spoiled? Many animals, like caribou, were easy targets because of their predictable habits and manageable size. Mammoth hunting, by itself, would not support the needs of a mobile, widely ranging people, and was probably not their main source of sustenance. At least some of the association of humans with mammoth remains may result from scavenging marrow for calories, and bone for tool blanks. The wide range of size in Clovis projectiles makes it likely that early people were highly versatile when it came to exploiting food resources. Regardless of the environment that Clovis people faced, they were favored by a highly diverse array of animal species, without large congregations of any one kind. It would have fostered generalization in procuring game rather than specialization. By not choosing to follow any particular game, they could have roamed freely and exploited whatever animals they encountered.

Wet times, that clearly prevailed for Folsom hunters, generated greater forage for bison that stepped into prominence after other grazing species disappeared, but available water also removed the need for animals to hold to patterned movements. When Folsom hunters had to forage for non-game resources, they would have been hard pressed to predict where the game might be relocated. Today's cattle tend to congregate at windmill pumps. How scattered might they become if there were ponds everywhere? The end of Folsom times coincides with an abrupt and drastic shift to arid conditions that would have forced bison to form herds and move in somewhat predictable patterns during their frequent visits to the relatively few water sources. Knowledge of animal behavior works much to a hunter's advantage since, among other strategies, one can wait in ambush along established routes.

What caused fluting technology to disappear?

Evidence at the Gault Clovis site, north of Austin, Texas, shows a lifestyle that persisted long after Pleistocene fauna disappeared (Collins, 2002). Over thousands of years, patterns including functionality of camp tools and use of certain parts of the landscape, are essentially constant. Changes in point styles could make a good argument that people had been displaced again and again over time, but all the other site evidence makes a case that the same people adopted a succession of new technologies to utilize an inventory of local resources that remained unchanged for centuries.

When climatic stability returned, strategies were developed to fit locally specific needs. As in other parts of the world, when the ecology remains stable, so does the pattern of its exploitation. While fluting technology was a brilliant solution to deal with a desperate catastrophe, it was not suitable once the crisis ended. Cultural remnants at the Jones-Miller site in Eastern Colorado (Stanford, 1978) indicate that hunters using Hell Gap projectile point styles, 9,000 years ago, could have shared a wide array of sociological traits with much later buffalo hunters of the northern plains that were documented by early explorers. The hunt apparently depended on luring the animals by exploiting knowledge of bison behavior. Artifacts at the kill site are reminiscent of ritual practices observed in historic hunts. In sum, despite many differences in the appearance of their tool kits and projectile points, the lifestyle seemed to represent a 9,000-year continuity.

Although fluting can be relatively risk-free when edge-to-edge flaking reinforces knapping skill, it would be difficult if bifacing were to be done differently. Therefore, changing requirements for tool flakes could easily be the most important factor in the abandonment of fluting. When people found that they could sustain themselves comfortably without having to travel so widely, the trade-off for relaxing maximized flake utility was the necessary abandonment of fluting.

Conclusions

Fluting traditions chronicle human adaptation to relatively drastic shifts in ecology. The strongest shifts in technology have coincided with the greatest shifts in nature. Instead of being displaced though, people simply adapted their technology to be more in tune with what nature threw their way. There are indications that point styles leading to Clovis fluting also transformed into other styles than Clovis. Styles prior to fluting were not distinctive enough to draw attention without a way to date them. Not only was the pioneer population sparse, thirteen thousand years of subsequent activity have effectively masked their presence. Edge-to-edge flaking allowed people to stay away from stone sources for longer times, and incidentally led to fluting. So, when the need for mobility declined, the practice of fluting disappeared as well.

Chapter 10

Replication

No matter how well archaeologists collect their data and theorize about how stone tools were made, the most persuasive evidence comes from actual hands-on duplication of original processes. Scientific replication requires an endless cycle of hypothesis, trial, and comparison with artifacts at every stage. By endless, I mean that some variables may be undecipherable. For example, it is known that water weakens the atomic bonds of silica, making stone easier to break, and caches of preforms have been found soaking in springs, but since the effect is not permanent, there is no way to tell how much the stone was altered for the original knapper. As our knowledge of fracture physics improves, there will be more opportunities to do computer modeling of the pertinent variables and their relationships to one another. Ultimately though, it is up to the flintknapper to demonstrate which actions provide specified results, and which do not.

Anyone who has spent time around flintknappers knows that they develop allegiance to particular approaches based on their success or familiarity with such techniques. Divergent approaches can be defended or detracted depending on whether one thinks the goal is replication, art, or commercialism. Because flutes can be made by a wide array of methods, it has been difficult to come to agreement on what was actually done in the past.

Most lithic technologists are content that rudimentary flutes which feather out after an inch or two can be made by the same percussion techniques used to create bifaces. That finding is consistent with the fluting lineage we have just proposed. However, as longer and longer flutes are attempted, success is more difficult to achieve. Modern knappers often try to control the most troublesome variables by using techniques borrowed from their modern surroundings. Thus, leverage mechanisms take over the delivery of force, and durable copper replaces antler. If preforms are judged to move too much, they are held in place with vises and clamps. Often, early stages of preform shaping are achieved by grinding rather than flaking. But can these substitutions of convenience properly explain how people removed long flutes over 10,000 years ago?

Ancient craftsmen were tying to maintain effective weaponry that accommodated their hafting system. That is very different from simply reproducing the appearance of a point. To mimic an end result fails to demonstrate its incremental development. To make the right tool on demand requires that each of the necessary production steps to be applied correctly, and in the right order. In other words, the process must be managed. As we learned earlier, traditions captured successful innovations bit by bit and kept them stable. Because each accommodation was relatively minor, there was little risk of failure. It would be foolhardy to argue that early people lacked the capability to accomplish any process that modern knappers use. However, they were likely motivated to do as much as possible with pre-existing skills rather than come up with brand new ones each time change was required. The most promising methods will be those that provide an audit trail of how one knapping process could be transformed incrementally to the next.

Evaluation criteria

Fluting has now been successfully demonstrated by such a wide range of methods that we need to examine them against a set of criteria to see what ancient knappers would have done. All the methods have their positive attributes, but some do not correlate well with the past. My criteria are designed to address the feasibility for fluting under the conditions prevailing over 10,000-years ago, when there were no convenient means of transport or ready access to exotic construction materials.

Since no single knapper has become expert in all of these disparate fluting methods, much of the discussion must rely on reports by experimenters who have a naturally biased interest in promoting their own technique. Objective criteria can help resolve this understandable tendency.

Clearly, correlation with archaeological evidence greatly improves the credibility of a proposal. From what we know about traditions, we should expect their technologies to be internally consistent. The resources required to accomplish a fluting method must also be sufficiently available to permit routine use. Acknowledging the mobility associated with fluting causes us to consider how knapping tools might facilitate or impede peoples movement. Related to that concern is how easily a flute can be produced, and how reliable the process.

Compatibility with archaeological evidence

In general, the technological trend through time has been from percussion to pressure techniques, although percussion is invariably used at an early stage of any process. As we have seen, the rebound of percussion advances fracture much faster than pressure allows. That difference in physics leads to considerably different knapping practice that leaves distinctive flake scars.

Characteristic details of flute scars provide valuable clues to suggest the approach used for fluting in ancient times. Attributes of flute scars, like spirals, arced flake trajectories, wavers, and irregular scar edges, indicate when the preform was weakly supported. Flatness depends on stability of support, as well as the speed of fracture. Flake terminations can indicate whether pressure, punch, or percussion techniques were used. For example, hinge flakes are highly unlikely when using pressure, and shallow steps almost never accompany percussion.

Ideally, each of the remaining criteria should be supported by direct physical evidence from the archaeological record. Not only is it important to match results from modern methods against ancient results, we have to know the range of actions that can produce those results.

Technologies consistent within the tradition

Apart from being able to replicate a flute scar, the fluting technique must fit in with the larger technology base, including camp tools. If leverage was crucial for fluting, shouldn't it have been applied to solve other technical problems? Similarly, the precision fits required to construct leverage devices would have encouraged carving skills that should be visible archaeologically.

Each of the proposed methods for taking a flute requires very different knapping habits. It is reasonable to expect that skills required by a proposed method of fluting should be expressed in other aspects of the craft. Why should mechanically aided pressure fluting not lead to making preforms by the same means? Likewise, fluting by punch would be incongruous with making preforms by direct percussion. Therefore, as fluting is taken to extremes of length and thinness, we must keep asking if it can be viewed as an integrated, holistic system, or if special variations might indeed have risen spontaneously.

Compatibility with available resources

Each method of fluting works best with stone of a particular quality. Heat-treating makes stone brittle and easier to flake, but that same brittleness makes the resulting artifact more vulnerable to damage. A fluting method that requires heat treating to succeed necessarily produces a more fragile point. Improving the ease of manufacture in this way not only risks failure of a projectile, it lessens the life of tool flakes.

Not all fluting methods deal convincingly with the same range of toughness that is found archaeologically. Modern knappers routinely use exotic materials that were not locally available in ancient times. Leverage systems require components with unusual degrees of tensile and compressive strength, flexibility, and resistance to fatigue. Providing those components equally throughout the range of fluting practice may not have been possible.

Portability

A knapping tool kit must perform reliably during long treks through a variety of ecosystems. Would the kit have been easy to maintain and transport? The answer depends in part on whether construction would be simple or complex. Complex mechanical fluting devices are tricky to calibrate and cumbersome to carry, which makes them generally unsuitable for a mobile lifestyle. When fluting is possible with a minimal tool set, how can extra equipment be justified?

Reliability and ease of use

A technology is most robust when process controls bypass the need for individual virtuosity. How difficult or easy each method is to master influences how well it could have been transferred from generation to generation. This is perhaps the least objective criterion that can be brought to bear, and one that is fiercely contested because each experimenter inevitably has more experience in one approach than another. Whatever the means chosen for fluting, routine success requires that every step of the process be controlled. Conditions that produce a desired result can either be tolerant of variation, or extremely unforgiving. If the range of tolerance is too tight, it increases the prospect of failure in a process step. Unless the process controls provide sufficiently consistent results, replicas will not match archaeological samples.

Logically, greater success should come from utilizing similar methods for a wide range of processes than from using divergent approaches. A knapper's muscles are trained to bear down at an angle and loading rate that consistently produces the right kind of flake. Switching from one style to another makes predictable flakes difficult unless new process controls are added. Even though process controls are indispensable for fluting, the more controls a knapper has to remember, the longer it takes to become proficient.

Fluting methods

Physics dictates the most obvious differences between fluting by pressure and by percussion. Impact allows greater levels of potential energy to be stored, which leads to faster and smoother fracture advance. It also enjoys the greatest latitude of blow angle and contact time with the hammer. The physics of impact makes it possible to create much larger flakes by percussion than can be achieved by an equivalent exertion of unaided pressure. That makes percussion the most reasonable foundation from which subsequent techniques for fluting could be directly developed. Punches should be considered a form of percussion that permits the accuracy of pressure.

Pressure flaking requires a more sturdy platform than would be needed to produce the identical flake by percussion. A platform prepared to aboriginal fluting standards will usually collapse before enough potential energy can be built to run a full-length flake. The amount of potential energy available at the start of the flake determines the speed of fracture propagation. Therefore, percussion flakes travel much faster than those made by pressure and consequently are less affected by surface irregularities or defects in the stone.

Pressure fluting encourages long preforms because their extra flexibility increases available contact time. It is less useful on short preforms. The end-to-end compression inevitable with pressure fluting forces the flute along a path between the platform and the tip support. Therefore, feathered flutes of appreciable length are especially difficult to obtain. Early Clovis fluting exhibits feathered scars too large to have been made by unaided pressure.

Computer modeling of fracture predicts that hinge terminations may only be made by percussion. If hinge terminations are detected on archeological fluted points, that can be regarded as unequivocal evidence they were not made by pressure. However, lack of hinges may not be a useful diagnostic if aboriginal knappers favored stone of a quality that precluded hinges.

To my knowledge, no compelling rationale has been advanced to bridge the developmental gap between percussion and pressure methods of fluting. Through known history, pressure has been reserved for edge refinement and patterning of shallow scars, while percussion has been preferred for large flake removals. Also, it is abundantly clear that percussion techniques must be mastered in order to create tool flakes and shape the preform in preparation for *any* means of fluting.

Attempts to replicate fluting have led many skilled knappers to confirm the viability of a wide variety of methods. Rather than address each one individually, I have grouped them into broad categories based on principles of physics and differences in equipment. Each proposed solution demonstrates mastery of one the most difficult challenges presented by ancient technology. Failure to match archaeological materials does not epitomize the competence of the investigator.

Compound leverage

By far, the most popular modern method for fluting points is compound leverage, because it provides exquisite control on force delivered to the preform (Solberger, 1985). A compound fluting jig approaches the mathematical force vector used in computer modeling and offers to be a valuable source of data for testing predictions of the computer model.

Leverage fluting is impressive when making large flutes, but is increasingly difficult to use as the preform length declines. Each length of preform, even for a second flute, requires adjustments of the lever through shims and notches or screws. Flute trajectories typically arc from end-to-end, which severely limits the capacity for truly thin fluted points. Changing to a new paradigm for preform design might avoid arced flutes, but thin preforms are still likely to bend and break under leverage. A related problem is that most replicators build preforms in a way that anticipates very little retouch after the flute removal, quite different from how most archaeological specimens were made. Most applications of compound leverage require an edge clamp high on the preform, which leads to early overshots more than to tip overshots. Using vises or clamps to support fluting preforms should leave at least some damaged edges, but delicate edges on preform artifacts are unblemished.

Leverage

A

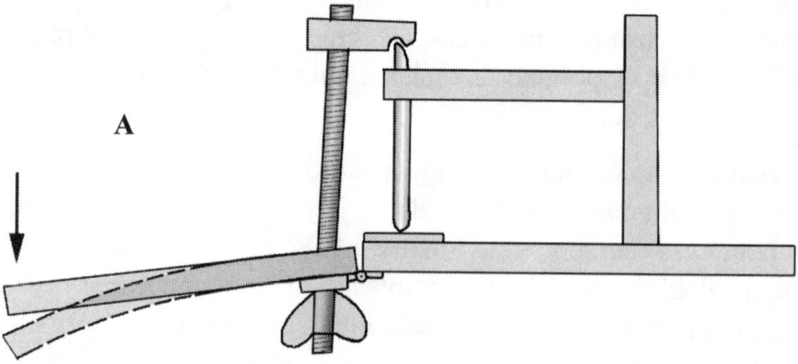

B

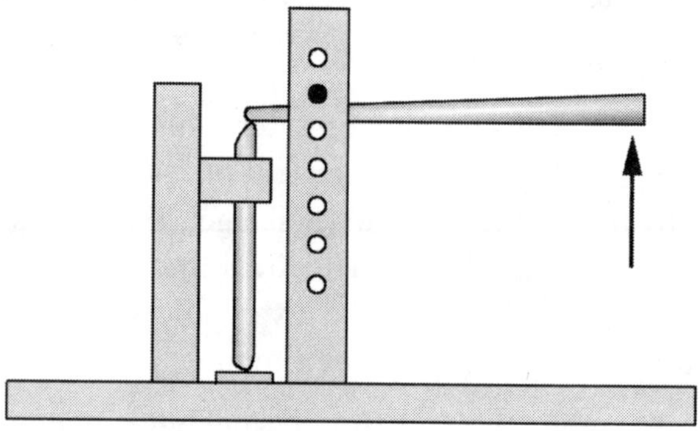

To make long, smooth flutes, a strong platform must store sufficient potential energy in both the lever and tension element.

A. Solberger-style compound leverage
B. Direct leverage

The mechanical advantage that generates enough force to drive a flute is also enough to break components made of wood or bone. Because the strength of raw stone can quickly overpower weak mechanical components, most replicators use metal leverage arms, heat treat their preforms, and build non-aboriginal style platforms strong enough to resist the stress. Constructing such a device in a primitive context requires precision notching, grooving and shaping that is not manifested in other archaeologically recovered tools of the period.

Direct leverage

Construction of a direct leverage device is relatively simple, but it takes greater training to achieve consistent results because cam-like toggling between the lever and the preform makes it difficult to control the angle in which force is directed. Experiments show that relatively compact equipment can be designed, but any misalignment tends to be amplified.

Both direct and compound leverage rely on advanced mechanical principles that are not apparent in other phases of the fluting tradition. Precision fits of highly durable components required by both methods should have led to joinery that isn't seen archaeologically. The durability required for leverage devices would have required specialized production, the carrying of cumbersome equipment, and a high labor cost to maintain the jig. The fact that many original preforms have edges too delicate, sharp, or uneven to be clamped also makes aboriginal use of leverage devices difficult to accept.

Axial compression

A self-taught knapper, Swoose Alexander, devised a unique system for fluting that uses axial pressure, and bypasses the need for an edge clamp. He binds the ends of a forked branch to put the preform under tremendous end-to-end pressure and initiates the flute by a sharp lateral tap near the preform base.

Aside from the initiating tap, this system only provides an axial load. Because there is no opportunity to create critical bending stresses, this system can only apply to full-length fluting, and would avoid overshot terminations found in the archeological record. A subtle objection to this method is that fractures cannot stop on course, while Folsom flutes commonly end abruptly just prior to the tip. Splits or burinations would be rare, and developmental steps from earlier fractional flutes are unexplained. It remains to be demonstrated whether relatively delicate Folsom preforms can stand up to the extreme compression loads. The time and effort demanded for fabricating argue that it would need to be transported rather than build a new device each time fluting was required.

Swoose uses a method of pressure flaking by extreme compression that sends flakes from edge-to-edge, and bears no resemblance to the style used to prepare aboriginal preforms for fluting. By utilizing his experience with pressure flaking to arrive at a feasible method for fluting, he demonstrates the kind of developmental sequence we might expect to support his concept archaeologically. Yet, the lack of material evidence representing a transition to axial loading argues against fluting by any mode of pressure. Had Swoose been accustomed to another style of pressure flaking, it is less likely that he would have arrived at the same solution to fluting.

Unaided pressure

To my knowledge, only Gene Gryba (1988), of Calgary, Alberta, has replicated Folsom fluting by direct pressure. Apart from the standard set of tools required for initial percussion and pressure refinement, Gene uses small antler rods the length of his forefinger. Not only is this the most minimal tool set proposed for pressure, it matches the simplest tools for direct percussion. A man of average build, Gene, can force off a four-inch flute by pressure—a feat of strength that few even attempt to duplicate. The preform is held in the fist of one hand while the bone rod is forced against the platform by the heel of the other hand. Both hands are supported between the knappers legs, which provides additional pressure. While an impressive magnitude of force is brought to bear, high-grade or heat-treated stone is required to remove normal-size flutes by hand, indicating that the technique is at the upper limit of its capacity.

This technique does not require anvil support, and therefore does not explain why tips of archaeological preforms were ground. Because the lateral support is weak, channel flakes tend to spiral and undulate more than archaeological examples, which would cause problems with hafting.

Unaided direct pressure

Removing flutes by unaided direct pressure requires a knapper to combine his legs, arms and hands to generate sufficient force.

Crutch pressure

Crutch pressure techniques have been most extensively explored to replicate Mesoamerican obsidian blade production (Crabtree 1966). The presser is attached to a long stick, and body weight is used to generate enough force to separate the blade from a core. One advantage of the crutch approach is that it effectively increases torque at the tool tip and thus represents a larger angle of blow than do some other pressure methods. Experiments show that while the approach works rather well for glassy materials, it is difficult to flute tougher grades of stone, or preforms exceeding two inches in length. Built like an oversized "T," a crutch can store no more potential energy than the knapper's body weight allows.

Although the pressure crutch is viewed by some as an extension of an anthropologically observed technique, anthropological accounts are so vague that we don't know for sure how blades were made. There is no indication that fluting traditions would have depended on such an elaborate technique that requires great strength, special material quality, and specialized equipment when people were constantly on the move. There is certainly no evidence that Mesoamerican blade techniques rose from earlier fluting traditions, and no indication that the crutch played any role in the making of ordinary stone implements for day-to-day use.

Crutch pressure

Direct punch

Following the principle of Newton's Cradle (page 83), a punch can transfer the blow of a hammerstone precisely to a specific impact site. Transferring energy through the intermediate punch makes it very difficult to achieve proper balance between support and hammer, and can waste energy. Because the hammerstone does not touch the preform, the punch is the only component that is sensitive to serious damage. But punches made of antler are easily damaged because the blow is aligned with the grain of the tool.

An interesting method for making very long stone blades occurred in Europe between about 2,400 BC and 2,900 BC. It used a set of process controls nearly identical to those described for Folsom projectile points. The cores are called "pound of butter" cores because they resemble an old style of butter packaging. Based on the consistent location of impacts on arrises, every flake on a pound-of-butter core appears to have been made by punch. The resulting blades, between 24 and 40 cm long and only 5 to 8 mm thick, were exported as specialized trade items rather than used as everyday tools, while the cores and shaping flakes were discarded at the workshop. The fact that techniques were not co-mingled fits with earlier speculation that people are not likely to force additional requirements on themselves when a single knapping method can be used. Proportions are quite near to

Pound of butter core

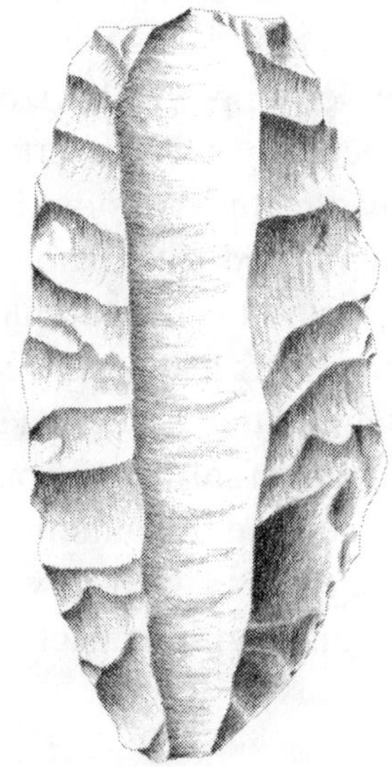

Up to eighteen inches long, these cores from Europe were used only to produce blades that were traded for other goods. Made entirely by punch, they are proportioned very much like Folsom points.

those of Folsom preforms, except that the cores are thicker, and the shapes are strikingly similar as well. These two technologies are far removed from each other in time, space, scale, and intent, but can be explained by the same principles of physics. The fact that, after scale is taken into consideration, blades of comparable delicacy to Folsom channel flakes were made by punching shows that punching has to be included in the array of possible methods for fluting.

Rocker punch

To overcome the problem of splitting direct punches, rocker punches are inclined, to resist the blow where the tool strength is greatest (Patten, 1999; Geib and Ahler 2002:249-271). Tool placement is precise, and the rocker is hit near its tip to deliver the force for fluting. There are many possible variations with a rocker system. Support can range from using the knapper's feet as a vise, to pegs, or the use of stone rests, and can involve mechanical components or another knapper. Even the springiness of the ground can be significant because sand, forest duff, or caliche hardpan all offer different levels of resistance. In practice, the variability of support makes standard results unlikely. I have achieved excellent matches to archeological flute scars with controlled lab conditions, but typically ruin my preforms when I go "on the road."

Other knappers have modified my original rocker punch concept to operate through a wide range of biface reduction requirements, from tool flakes to biface refinement. Using a short section of antler supported horizontally on a leather pad by the knapper's heel allows flakes virtually equivalent to what can be accomplished by direct percussion. Even Clovis biface core flakes are easily duplicated with simple tools. The key is to select a hammer that conveys the right impulse through the punch. Substituting stone for antler lets a knapper reduce large quarry chunks with ease by the same action used on a small biface. Perhaps the most impressive aspect of this rocker punch approach is that the motor skills can be developed in hours instead of days or months. While it sounds that the rocker punch serves every need, it should leave more evidence of off-margin impacts than is observed.

Rocker punch

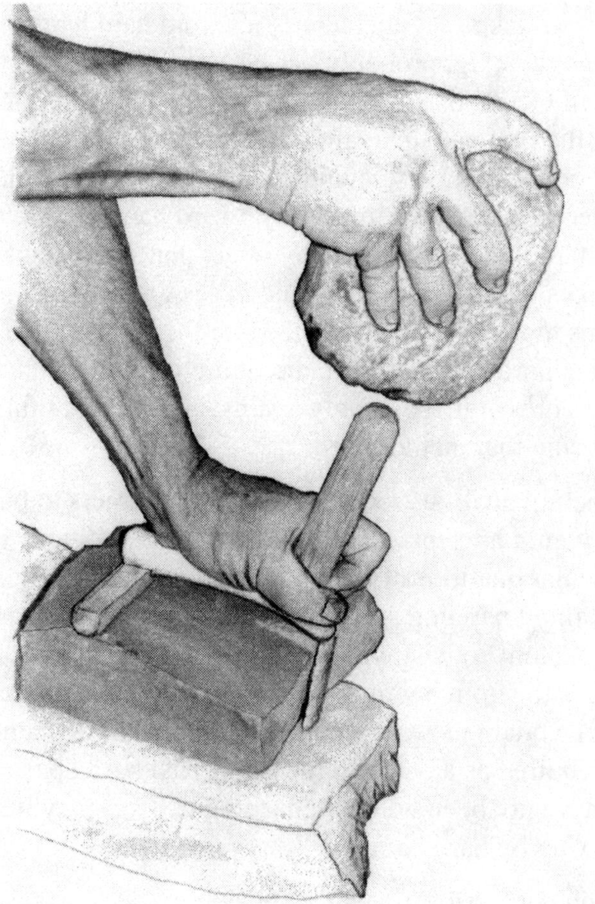

This method of fluting produces good results, but is overly sensitive to changes in condition when moving from place to place. Unless each of several components is tuned to the others, failure will occur.

The tool kit required for a rocker-punch is universally available and pocket sized. A small section of antler can be used as long as it is hit with the right-sized hammer. Rocks normally unsuitable for knapping hammers could be employed, providing that the punch is strong enough to take impact with the core. To date, though, no one has demonstrated every phase of aboriginal knapping by rocker-punch.

Hard hammer percussion

At the first Folsom workshop in Austin (Clark and Collins, 2002), experts judged flake scar characteristics from hard hammer to be especially convincing, probably because of the dynamic mode of fracture. Phil Geib (Ahler and Geib 1997) had moderate success by supporting the preform tip against an anvil, but reported needing "Zen-like" concentration to avoid failure. Few delicate platforms can be expected to survive impact by a hard hammer, although archaeology provides us with many intact platform remnants on channel flakes. Hard hammers would necessarily make hinge terminations more likely because the hammer is so stiff. The lack of distinctive hard hammer signature initiations on archaeological bifaces and cores argues strongly against aboriginal fluting traditions using that method.

The exacting qualities required for hard hammers to be suitable for fluting would have placed undue burden on people. Rocks with those particular qualities of toughness and grit are nearly as scarce as good quality knapping stone. Sources of good hammerstones for fluting seldom coexist with those of good knapping stone, and are different still from hammers needed to section large material. If hammerstone qualities were so critical, extra hammerstones would surely be required as a back-up measure. That such special hard hammers have not been noted archaeologically greatly weakens the case for fluting by hard hammer.

Antler baton percussion

Experimental replication proves that direct percussion by antler can duplicate archaeological results in both projectiles and camp flakes, without needing separate knapping tools (Patten 2002: 299-308). I support a short section of antler against my upper calf muscle, and hold the preform vertically. The blow delivery is relaxed, to best establish rebound equilibrium between hammer and core. No extra equipment is needed beyond that normally employed to make bifacial tools.

Terminations of flutes made by baton percussion often appear to turn slightly into the body of the preform, an attribute characteristic of authentic Folsom flute scars that has not been observed with any other method of fluting. This termination always occurs very close

Direct percussion with hard hammer

In this instance, the preform is held horizontally against a vertical wooden anvil, while a small sandstone hammer is swept downwards, using a glancing blow.

to the preform tip, and allows the scar trajectory to remain flat. Other methods of fluting tend to overshoot when the flute extends so near the tip, and overshots normally allow the channel scar to arc more than is observed in ancient artifacts. This subtle clue may be the best direct evidence available that baton percussion was favored for fluting.

Because preforms tend to fail in overshot flakes near the point of support, bend breaks are diagnostic of where the preform is supported. Therefore, a relatively large number of archaeological overshots near the tip indicate that preforms were routinely supported below the center of the preform. Collateral damage like shattered ears, split or burinated preforms are also most easily explained by direct percussion.

Direct percussion with antler

If the preform is positioned against an anvil so the blow is the same as used for normal bifacing, fluting can be accomplished as easily and reliably as for any other flake

Evaluation

Now that we have an idea of what each fluting technique involves, we can look at how they stand up to the evaluation criteria proposed earlier.

Compatibility with archaeological evidence

Given that there is no conclusive direct evidence of fluting methods, we need to be most critical when expected clues are missing, and when well-documented clues are not accounted for. For example, unaided pressure should not require ground preform tips that are so common archaeologically, and shaped components from leverage devices are conspicuously absent.

Leverage methods cannot produce flute scars of appropriate quality without the most extreme preparation of the preform surface, because a static load is most susceptible to bending that distorts the flake path. Percussive modes of fluting, however, are capable of leaving clean, flat flake scars on even a poorly prepared preform. When the quality of the flute scar is considered, few modern replications can match the original work on a consistent basis. The difference is more than aesthetic, it cuts to very heart of functionality. Without that quality of scar, it is impossible to meet dimensional standards, and it becomes difficult to fit the projectile to the haft unless special accommodation is made. Direct percussion seems best suited to match archaeological results.

Technologies consistent within the tradition

The biggest problem with methods of fluting other than direct percussion by antler baton is that they fail to demonstrate how they would have come into use. Simply looking at how few stoneworking traditions around the world routinely used longitudinal basal thinning shows how unlikely it was for fluting to develop in the first place. Alternative fluting methods are all subject to enough technical problems that the difficulty of coordinating the proper percussion blow seems trivial. After all, it takes the same accuracy to make any competent biface. Why would people complicate their technology when a simple expedient was already imbedded in their normal stoneworking process? Each successive modification of fluting practice should be explainable as a logical extension of prior practice. Most of the proposed innovations require new forms of technology, like precision fitting, that should impact other everyday tasks, and we see no such development.

Pressure, like percussion, is normally mastered by stages of increasing difficulty. Almost anyone can nibble edges to sharpen them or prepare platforms for percussion. Smoothing facets by taking flakes that at least reach midline is considerably more difficult. And thinning by edge-to-edge flakes is generally not seen until after the practice of fluting was abandoned. Without evidence for an intermediate level of pressure flaking, the leap to fluting by pressure seems unlikely.

If punches were a standard knapping tool, the requirement for an extra tool component is not serious of itself, but it should cause us to question what other uses of punches might occur. Using punches for fluting would only make sense if the punch were also used for reducing the core. It is more convenient to make blades by the punch method than to punch out bifacial flakes as the basis for camp tools. Had the microblade tradition preceded fluting, projectiles could have been made from large blades far easier than by bifacing. It is difficult to imagine a blade-based technology converting to bifacing, then reverting to fluting by blade technology.

Compatibility with available resources

An easily overlooked objection to non-percussion fluting methods is that they place undue emphasis on making the projectile point when, as we have seen, the real value was in acquiring useful camp tool flakes. Focusing on the finished point has led many replicators to match projectiles at the end of their useful life, rather than at the beginning. The failure to account for the production of other tools comes from thinking of projectile points as a goal independent from all others. Percussion fluting provides a seamless interface between making camp tools and refining the projectile.

Portability

Methods of fluting that require special preparation or extra parts to manufacture, maintain, or carry would put an unnecessary encumbrance on the group. Without beasts of burden, it is certain that early nomads were accustomed to carrying significant loads day-by-day. However, accounting for backup equipment to guard against failure, and the fact that most adults needed ready access to the knapping kit, a light tool set must have been valued. Most pressure methods add time and labor for fabricating components that must be calibrated to work effectively. Once that much effort has been expended, it would only be reasonable to maintain and transport the gear in order to minimize set-up cost each time a replacement projectile is needed.

Only unaided pressure and direct percussion require no additional equipment beyond that necessary for making any other bifacial tool.

Reliability and ease of use

While I acknowledge that I am most familiar with fluting by an antler baton, percussion is a skill familiar to any stone-age people. The reliability of fluting by percussion is directly correlated to the reliability of creating camp tool flakes required by any other tradition as well. In my experience, mechanical solutions are prone to unexpected variations when components are replaced with materials of even slightly different dimensions or physical properties. Any technique that demands manipulation of unfamiliar tools risks failure.

Conclusions

The rise and fall of fluting traditions can be easily explained by a reasonable chain of events using direct percussion with antler tools. Percussion best satisfies the requirements of long contact duration and high fracture speed that is responsible for the unique appearance of fluted points. The requirement to use the same stroke, whether shaping a biface or fluting it, is speculative, but greatly reduces the training needed to be proficient in what appears to be a risky task.

Whether ancient people fluted their projectiles by only one means or by many is important to our interpretation of their behavior. Concluding that a single method was preferred when a multitude of options were theoretically available leads us to deeper questions about why the choice was made. I do not doubt people of an earlier time possessed the inventive capacity to imagine any technique we can dream up today. It is much more likely that they weighed the advantages and disadvantages of many possibilities, but stayed their course for practical reasons. Replication happens to be one of the best means by which we can appreciate their decision-making processes.

Chapter 11

Concluding Remarks

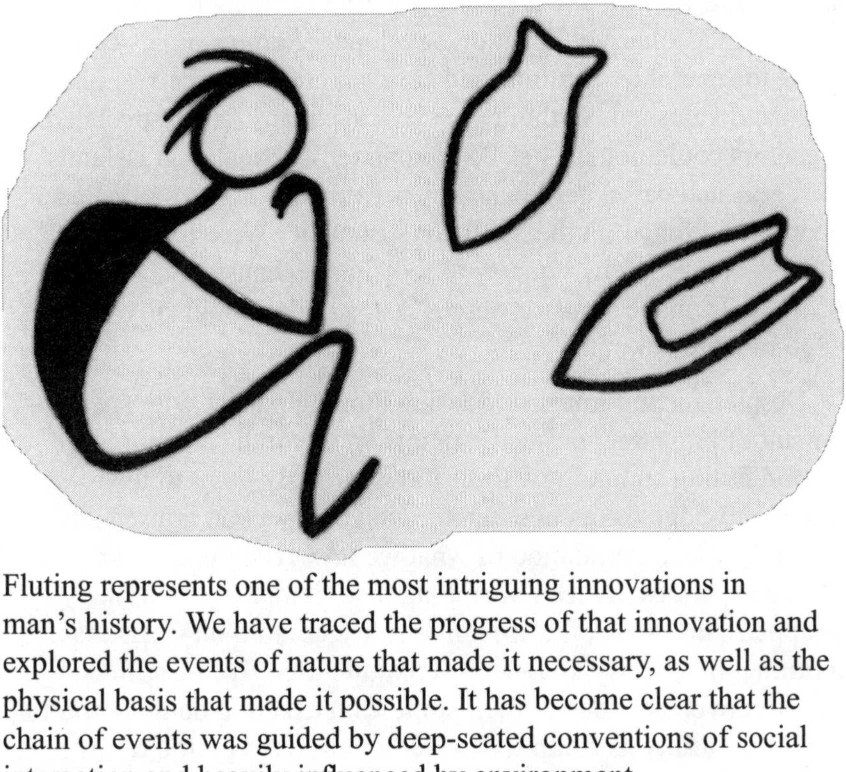

Fluting represents one of the most intriguing innovations in man's history. We have traced the progress of that innovation and explored the events of nature that made it necessary, as well as the physical basis that made it possible. It has become clear that the chain of events was guided by deep-seated conventions of social interaction and heavily influenced by environment.

No matter how the ancestors of Clovis people got to the Americas, much of the land they saw at the end of the Pleistocene was a cool, Savannah-like expanse of tall grasses. Diverse species of animals grazed in widely spread herds, unlike the massive concentrations of bison encountered by early explorers. As long as the ice sheet was in place to buffer temperature extremes, ecological zones intermingled in ways that today are hard to imagine. Hunting pressure by small bands of people probably kept game, and people, on the move. If those early people stayed near watercourses, as would seem natural, most of their material leavings would either be eroded or buried under 13,500 years of sedimentation. If they relied on the bone of mega-beasts as a source of tool material, their presence would be further concealed because after so many thousands of years bone reverts to dust.

The shift to the Holocene occurred in little over forty years, and it left a very different ecology. Where ecological zones had intermingled, sharp delineation developed. Some animals could move to favorable locations and survive, but others had no place to go and vanished. Without prey species, many large, specialized predators could not subsist. As rain patterns irrigated grasslands on a seasonal basis, herd behavior became more migratory, and people who relied on the herds for sustenance were forced to adapt as well. Returning to a quarry to replenish exhausted stone could be risky. To make stone resources last, strategies logically included edge-to-edge flaking.

Despite stirring images of daring hunting parties with spears in pursuit of big game, projectile points were *not* the driving force behind fluting technology. Indications strongly point to the role of edge-to-edge flakes in extending the utility of an individual rock as the true foundation of what we now recognize as the fluting traditions. Instead of ripping their compactly "shrink-wrapped" stone package apart like an eager child on Christmas morning, these early knappers were quite literally "skin flints," who removed portions of their stone with exacting deliberation to maximize utility as camp tools over a protracted period of time. The edge-to-edge flaking concept led naturally enough to the fluted projectile point that has been recognized as one of the outstanding achievements of human ingenuity. However, the projectile could have been made in any number of forms to do its simple job, as Solutrian hunters aptly demonstrated thousands of years earlier with a very similar technology, sans flutes.

We are coming to recognize fluted points as the special prize found in the deepest recesses of a quarry blank. The modern hunter who has his gun stock checkered and barrel engraved may be in the best position to understand why fluted points came to be. Only an attentive craftsmen could turn out a fluted point in its full glory, and a well-executed implement encourages competent use. Over time, the expression of fluting ideals took different forms in various settings, but lost out to other techniques for making tools as settlement forced a retreat from broadly mobile strategies and encouraged greater generalization.

Refocusing our attention from projectile points to mundane camp tools may not satisfy our sense of the dramatic, but it is necessary for a proper appreciation of how original practitioners valued components of technology in relative terms. Piece by piece, camp tools rarely excite a collector's enthusiasm, but in bulk, they dominate the archaeological record. The universal nature of camp tools makes a good case for their great importance as indispensable items of survival. Subtle differences in the overall process of acquiring and dismantling stone resources need to be studied in detail to understand their impact on adaptation to specific environments. Continued development of tool-use cycle concepts should allow better resolution of the relative roles and impacts of various activities. Transforming archaeological interpretation from isolated activities to holistic scope will present some of the most serious challenges to the new generation of archeologists.

This is not to suggest that there is no value in studying projectile points. Lithic technology is simply much broader than the mere construction of weapons. Once the universal need for camp tools has been met, the final conversion to projectile point is most heavily colored by social concerns. Those cultural responses appear as progressive alterations of previous processes. The evolutionary aspect of projectile point design through time not only provides a basis for typology, it carries the potential for relatively fine detailing of lithic technology lineage. Before that potential can be realized, however, a great deal of work needs to be done in outlining process recipes.

The question of where fluting technology originated is still open. European sources provide nearly all the requisite developmental avenues, but much research will be required for that path to the Americas to gain wide acceptance. Asian roots carry considerable weight, mostly due to proximity, but call for uncharacteristically severe and rapid adjustments from microblade or crude biface technology as people entered the new continent. Based on the evidence at hand, neither proposal can be ruled entirely in or out. No "feel-good" answer is supported by direct evidence and there are many uncertainties as to timing, much less placement, of ancestor technologies. The amount of work being done may soon lay these issues to rest.

Nothing motivates the march of technology more than changes in living conditions. Few periods of early human history, New World and Old, faced more dramatic change than at the end of the Pleistocene. Those climatic changes caused enormous shifts in animal distribution and behavior, which compelled humans to adapt their hunting strategies and lithic technology. Whether the adaptation to environment started at the same time the New World was first populated, or later, is not nearly so important as the nature of that adaptation. Continuing climatic disruptions led fluting traditions through an incremental evolution that culminated in full-length fluting designs. Variability in the shape and size of Clovis projectiles, highly dispersed sites, and access to multiple game species, all point to highly flexible, adaptive people, well suited for spreading across the continent.

It is my belief that fluting emerged from the loosely structured direct percussion of bifacial cores to become a highly specialized discipline. For a time, that specialization provided a stable technologic basis, but when game became more predictable and stone did not have to be conserved as effectively, camp tools could be made with less concern for waste. A shift of the base technology for utilized flakes does not necessarily mean that the culture changed as well. The means to supply camp tools is colored by routine access to resources. How the technology is exercised varies by distance from those sources. The greater the distance from replacement stone, the greater the inclination to rework the tool at hand rather than jettison it. In the time following fluting traditions, careful disassembly of transported stone was increasingly supplemented by casual reduction of local rock. Fluting could not persist without high standards of knapping, so more generalized lanceolate designs for projectile points were adopted to cope with the new environment. The apparent decline in skill does not imply that later knappers were any less competent, just that the emphasis on precision was relaxed. While hafting concerns may have led to fluting, there is no indication that hafting had anything to do with the cessation of fluting.

Since tool-flake technology supplies components that strongly support survival for the lifestyle to which the culture has adapted, it is driven by functional considerations that can change at a variable rate. Projectile design, on the other hand, responds in part to physics and stone availability, but is highly impacted by social values. Disconnecting the form of the projectile from camp tool requirements is important because it allows development of the two parts to be considered separately.

Lineage is revealed when we understand what attributes change with time. The shape of projectile points has long been a useful record of changing fashions—apart from strictly functional elements. But what appear to be otherwise meaningless successions in projectile design can be best explained by comparing process controls. Lack of a clearly supported pattern of development of lithic technology, and a poorly dated context of finds, has led to the lumping of too many fluted point styles into the category of "Clovis." Each level of advance in fluting required certain controls, and successive refinements decreed their own additional controls. It should be evident that full-length fluting styles, including Folsom, Gainey, Barnes, and Crowfield, are late developments in fluting traditions because they depend on the entire repertoire of process controls, including precision contouring of the preform surface. At this time, dated evidence is insufficient to set them in an unambiguous timeline.

Because fluting technology had to support normal everyday needs, competence in some level of stoneworking would likely have been common throughout all members of the society. We're not talking about the individual specialization required to produce stunning works of art like Egyptian or Danish daggers. But let us suppose that each member of an extended group could follow prescribed controls and make highly standardized products. That does not mean that everyone made Folsom points. But everyone must have been competent in making the basic tools needed to perform their own tasks. Such a broad base of competence is hard to equal in history, especially over such a long span of time. My replication experiments demonstrate that process controls allow reliable and predictable results without demanding individual virtuosity. That puts me at odds with many who insist that fluted points could only have been made by highly-skilled specialists. Though some level of skill is necessary, following a specific routine allows everyday workers to achieve impressive results. Stone tools made by later Paleo Indian traditions required just as much skill as fluting, but loosened process controls thwarted stylistic consistency that is often viewed as a measure of skill.

A combination of computer modeling, experimentation, and archaeological data allows us to conduct forensic evaluations of lithic scatters. Recognizing how to read flake scars as a measure of how the detaching force was applied gives archaeologists reliable data regarding knapping tools and physical behavior. The role of experimental flintknapping is strengthened when it is linked with physics. Once the principles of fracture have been deduced and verified, archaeological evidence can be interpreted directly, without having to rely on comparisons to modern knapping practice.

After considering the forensic evidence and the range of alternative ways in which fluting can be accomplished, I believe percussion fluting was the standard method used by *all* fluting traditions. We know that fluting can be accomplished by other means, but percussion is the only process whose attributes fit easily into an evolutionary scheme and match the archaeological data.

Earlier in the book I pointed to various issues that might have triggered the abandonment of fluting practice. In truth, the trigger could have been virtually anything. Catastrophic change inevitably arises from an accumulation of instability. Rather than accuse the final straw that broke the proverbial camel's back, we should look at each straw that contributed to the load. No single reason stands out. The inevitable abandonment came about because of changes in climate, settlement patterns, technological decisions, geographic range, available game, and any number of other contributing factors.

While fluting technology is outstanding in its appearance, it is just another state of equilibrium punctuated by the whim of nature. Although fluted points are so distinctive that they look like the work of people totally unlike anyone else, camp tools retain an underlying continuity. It is not until *after* Clovis times that territorial delineation of people becomes evident from the distribution of distinctive projectile styles that occur in restricted ranges. The relatively unimportant adaptations that took place while fluting was practiced is similar to the gradual fine-tuning of weaponry at any other time or place. The details may differ, but the human response always follows similar patterns. If cultural cohesiveness can be measured by technological similarities, then the people who shared the tradition of fluting appear to have grown apart with the passing of time.

When we apply a regional perspective, it becomes apparent that fluting traditions readily adapted to special conditions. The widely distributed resources of the plains made mobility the highest priority. Even when fluting was discontinued, the enforced mobility caused later Paleo Indian traditions to retain most of the earlier practices for building bifaces and maintaining stone tools. In the Great Basin, mountains defined self-contained resource pockets whose individual conditions varied widely. Clovis technology would have been difficult to practice without forced incursions from basin to basin. The variability in stone resources may have been instrumental in the rapid transformation to stemmed points, because much of the stone is unsuitable for fluting. At the Mockingbird Gap site, in south-central New Mexico, a broad geographic basin enclosed limited resources. The pattern of exploitation appears similar to that practiced by Archaic peoples of the Great Basin. Southeast of the high plains, where coastal plains met hilly uplands, inhabitants of the Gault site enjoyed plentiful stone, water, and a wide variety of edible resources. Their lifestyle was nearly indistinguishable from that practiced just prior to the arrival of Europeans. Farther east, copious stone sources led to greater activity at quarries, and less conservative practices afield than in the plains. If it were not for the unambiguous identifying trait of fluting, those groups might as easily be described as having adapted to cope with restricted regions.

As game migrations stabilized in response to drier conditions, successive generations would have found it easier to schedule tool stone forays without fear of losing track of game. More of their tool set could be prepared at the quarry, in advance of the need, diminishing their reliance on long-distance conservation of stone. As time progressed, stone supplemented from local and expedient sources accounted for more of the debitage at sites.

The unlikely chain of events set in place by the close of an ice age led people to use edge-to-edge flaking, and left them in possession of a fluting technology. Once established, that tradition transitioned in response to local conditions. Eventually, another climatic change triggered subsequent adaptation that led to the abandonment of fluting, without necessarily replacing one population by another. Even if invading populations existed, they could have adopted the strategies already proven satisfactory for survival. In other words, genetic inheritance is separate from technological practice.

Now the case rests, but it is not closed. There is no statute of limitations in archaeology, and each new piece of evidence or difference in interpretation can begin the process again. Sometimes new evidence drives new ways of seeing the problem, and speculation provides a new framework for interpreting evidence already gathered. Whatever the reason for re-examining the case, we have a responsibility to provide an audit trail that links successive arguments into defendable objective constructs. Unlike civil court, we should not be concerned with winning or losing the case. The goal is to uncover the truth, whatever it may be.

Bibliography

Anticipating the happy outcome that readers may want to delve further, bracketed annotations indicate sources that may open even broader perspectives. In any event, the most interesting reading may come from sources yet to exist.

Ahler, Stanley A. and Phil R. Geib

1997 *Considerations of Folsom Fluting by Direct Freehand Percussion.* Paper presented at the First Folsom Workshop, March 18-22, 1997. Austin, Texas.

Amick, Daniel S., editor

1999 *Folsom Lithic Technology: Explorations in Structure and Variation.* (Ann Arbor, International Monographs in Prehistory) [Eleven papers address a broad scope of Folsom activities.]

Anderson, David G. and Michael K. Faught

2000 "Paleoindian artifact distribution: evidence and implication" *Antiquity,* 74:507-13. [Geographic distribution of fluted points in North America.]

Andrefsky, William Jr.

1998 *Lithics: Macroscopic approaches to analysis.* (Cambridge, Cambridge University Press) [Outlines the traditional approach to lithic analysis.]

Baker, Tony

1998 *Application of Finite Element Analysis to the Understanding of Flake Formation.* ele.net/algor/ff_intro.htm

2000 "Digital Crabtree: Computer Simulation of Folsom Fluting." ele.net/algor/crabtree/crabintr.htm

2003 *A Theory for Flake Formation.* ele.net/algor/flake_creation/SD_setup.htm

Binford, Lewis R.

2001 *Constructing Frames of Reference: An analytical method for archaeological theory building using ethnographic and environmental data sets.* (Berkely and Los Angeles, University of California Press)

Bradley, Bruce and Dennis Stanford

2004 "The North Atlantic ice-edge corridor: a possible Paleolithic route to the New World", *World Archaeology*, 36(4):459-478. [An European alternative to Beringia.]

Bradley, Bruce A. and Evgenij Y. Giria

1996 Concepts of the Technological Analysis of Flaked Stone: A Case Study From The High Arctic. *Lithic Technology,* 21(1): 23-39.

Chandler, James M.

2001 "Immigrants from the Other Side?" Mammoth Trumpet, 17(1):11-16.

Clarke, Garry, David Leverington, James Teller and Arthur Dyke

2003 "Superlakes, Megafloods, and Abrupt Climate Change." *Science,* 301:922-923.

Clark, John E. and Michael B. Collins, editors

2002 *Folsom Technology and Lifeways*. (Special publication No. 4, Lithic Technology, Department of Anthropology, University of Tulsa, Tulsa) [Summary papers from the Folsom Workshop Conferences.]

Collins, Michael B.

1999 *Clovis Blade Technology*. (Austin, University of Texas Press).

Collins, Michael B.

2002 The Gault Site, Texas, and Clovis Research. *Athena Review,* 3(2) [Challenges Clovis-First Paradigm]

Cotterell, Brian and J. Kamminga

1987 "The Formation of Flakes." *American Antiquity,* 52:675-708. [Theoretical basis for understanding a limited range of fracture.]

Crabtree. Don E.

1966 "A Stoneworker's Approach to Analyzing and Replicating the Lindenmeier Folsom." in *TEBIWA, The Journal of the Idaho State University Museum,* 9:3-39. [Consideration of alternative fluting methods.]

1968 "Mesoamerican Polyhedral Cores and Prismatic Blades." *American Antiquity,* 33(4):446-478

Cullen, Ben Sandford

2000 *Contagious Ideas: on evolution, culture, archaeology, and Culture Virus Theory*. (Oxford, Oxbow Books) [Processes of cultural change.]

Davis, Loren G. and David A. Sisson

1998 "An Early Stemmed Point Cache from the Lower Salmon River Canyon of West-Central Idaho." *Current Research in the Pleistocene,* (15):12-14.

Dewar, Elaine

2001 *Bones: Discovering the First Americans.* (New York, Carroll & Graph Publishers) [The difficulties of reaching scientific consensus in a politically charged world.]

Diamond, Jared

1992 *The Third Chimpanzee: The Evolution and Future of the Human Animal.* (New York, Harper Perennial)

1999 *Guns, Germs, and Steel: The Fates of Human Societies*. (New York, W. W. Norton and Company) [The broad relationship between humans and their planet.]

Dillehay, Thomas D.

2001 *The Settlement of the Americas*. (New York, Basic Books) [A view from the Southern Hemisphere.]

Dixon, James

1993 *Quest for the Origins of the First Americans*. (Albuquerque, University of New

1999 *Bones, Boats, and Bison*. (Albuquerque, University of New Mexico Press).

Eaton, S. Boyd M.D., Marjorie Shostak, and Melvin Konner, M.D., Ph.D.

1988 *The Paleolithic Prescription.* (New York, Harper and Row) [Dietary requirements for primitive living.]

Firestone, Richard B.

2001 "Terrestrial Evidence of a Nuclear Catastrophe in Paleoindian Times." *Mammoth Trumpet,* 16(2):9-16.

Flenniken, Jeffery J.

1978 "Reevaluation of the Lindenmeier Folsom: A Replication Experiment in Lithic Technology." *American Antiquity*, 43:473-479. [Early attempt to reconcile archaeology with experimental results.]

Forrest, A.J.

1983 *Masters of Flint*. (Lavenham, Suffolk, Terrence Dalton Limited.)

Frison, George

1978 *Prehistoric Hunters of the High Plains.* (New York, Academic Press) [Excellent distillation of what archaeologists agree to be reasonably secure conclusions.]

Frison, George C.

2004 *Survival by Hunting: Prehistoric Human Predators and Animal Prey.* (Berkley, University of California Press) [The importance of understanding animal behavior.]

Frison, George C. and Bruce A. Bradley

1980 *Folsom Tools and Technology at the Hanson Site, Wyoming.* (Albuquerque, University of New Mexico Press) [Folsom workshop evidence.]

Frison, George and Bruce Bradley

1999 *The Fenn Cache Clovis Weapons and Tools.* (Santa Fe, One Horse Publishing Co.)

Geib Phil R. and Stanley A. Ahler

2002 "Considerations in Folsom Fluting and Evaluation of Hand Held Indirect Percussion." In *Folsom Technology and Lifeways*. (Special publication No. 4, Lithic Technology, Department of Anthropology, University of Tulsa, Tulsa) pp. 249:271.

Goebel, Ted, Michael R. Waters and Margarita Dikova

2003 "The Archaeology of Ushki Lake, Kamchatka, and the Pleistocene Peopling of the Americas." Science, 301:501-505.

Goodyear, Albert

2003 "The Stratigraphy Story at the Topper Site." *Mammoth Trumpet,* 18(3):12.

Gribbin, John and Mary

1990 *Children of the Ice: Climate and Human Origins.* (Cambridge, Basil Blackwell Ltd.) [Connects human development to changes in ice age climate.]

Gryba, Eugene M.

1988 "A Stone Age Method of Folsom Fluting." *Plains Anthropologist,* 33:53-66.

Guthrie, R. Dale

1990 *Frozen Fauna of the Mammoth Steppe: the story of Blue Babe.* (Chicago University of Chicago Press) [Detailed forensic account of lion predation on a Pleistocene Bison.]

Hall, Don Allen

1995 "The Tale of Two Sites." *Mammoth Trumpet.* 10(1):7

Hofman, J.L.

1995 "The Busse Cache: A Clovis-Age Find in Northwestern Kansas." *Current Research in the Pleistocene,* 12:17-18.

Holen, Steven R.

2001 *Clovis Mobility and Lithic Procurement on the Central Great Plains of North America.* Ph.D. dissertation. University of Kansas.

Holen, S.R. and D.W. May

2003 The La Sena and Shaffert Mammoth Sites: History of Investigations. In *Medicine Creek: Seventy Years of Archaeological Investigations*, edited by D. Roper, pp. 20-36. University of Alabama Press, Tuscaloosa

Hutchings, W. K.

1997 *The Paleoindian Fluted Point: Dart or Spear Armature? The Identification of Paleoindian Delivery Technology through the Analysis of Lithic Fracture Velocity*. (unpublished) Ph.D. dissertation, Department of Anthropology, Simon Fraser University, Vancouver.

Jodry, Margaret A.

1999 *Folsom Technological and Socioeconomic Strategies.* Ph.D. thesis, American University, Washington, D.C. [Addresses a broad scope of Folsom activities.]

Jodry, M.A., M.D. Turner, V. Spero, J.C. Turner, and D.J. Stanford

1996 "Folsom in the Colorado High Country." *Current Research in the Pleistocene,* 13:245-27

Judge, James

1973 *Paleoindian Occupation of the Central Rio Grande Valley in New Mexico*. (Albuquerque, University of New Mexico Press) [Based on the Ele Baker collection.]

Kelterborn, Peter

1981 "The Livre de Burre Method." *Flintknappers' Exchange,* 4(3):12-20

2002 "Measurable Flintknapping" in Experimentation and Interpretation in *Mesoamerican Lithic Technology*, Edited by Kenneth Hirth, University of Utah. pp.120-131. [Rigorous experiments to confirm theoretical predictions.]

Kunz, Michael, Michael Bever, and Constance Adkins

2003 "The Mesa Site: Paleoindians above the Arctic Circle." *BLM-Alaska Open File Report 86*.

Largent, Floyd B. Jr.

1991 "The Spatiotemporal Distribution and Characteristics of Folsom Projectile Points in Texas." *Plains Anthropologist,* 36(137):323-341.

LeTourneau, Phil

1997 Personal communication.

Morrow, Juliete E.

1995 "Clovis Projectile Manufacture: a perspective from the Ready/Lincoln Hills site." *Midcontinental Journal of Archaeology,* 20(2):167-191. [Clovis biface reduction in Illinois.]

Morrow, Julliete and Toby

1999 "Geographic Variation in Fluted Projectile Points: A Hemispheric Perspective." *American Antiquity,* 64(2):215-231.

O'Brien, Michael J. and R. Lee Lyman

1999 *Seriation, Stratigraphy, and Index Fossils: The Backbone of Archaeological Dating*. (New York, NY, Kluer Academic/ Plenum Publishers)

Patten, Bob

1999 *Old Tools-New Eyes: A primal primer of flintknapping.* (Denver, Stone Dagger Publications)

2002 "Solving the Folsom Fluting Problem." In *Folsom Technology and Lifeways*. (Special publication No. 4, Lithic Technology, Department of Anthropology, University of Tulsa, Tulsa) pp. 299-308.

Pelcin, Andrew

1996 *Controlled Experiments in the Production of Flake Attributes.* Ph.D. dissertation, Department of Anthropology, University of Pennsylvania, Philadelphia. [Empirical exploration of fracture paths in plate glass.]

Pielou, E. C.

1991 *After the Ice Age: The Return of Life to Glaciated North America.* (Chicago, University of Chicago Press) [18,000 years of ecological transformation.]

Pitulko, V.V., P.A. Nikolsky, E. Yu. Girya, A.E. Basilyan, V.E. Tumskoy, S.A. Koulakov, S.N. Astakhov, E.Yu. Pavlova and M.A. Anisimov

2004 "The Yana RHS Site: Humans in the Arctic Before the Last Glacial Maximum." *Science,* 303:52-56.

Prentiss, William C. and James C. Chatters

2003 "Cultural Diversification and Decimation in the Prehistoric Record" *Current Anthropology,* 44(1). [Why cultural changes progress at different rates.]

Quesada, Carmen Cacho, Jesus Jorda Pardo, Igacio de la Torre Sainz, Josde Yravedra Sainz de los Terreros

2001 "Tossal de la Roca (Alicante). New Data on the Mediterranean Magdalenian of the Iberian Peninsula." in *Trabajos de Prehistorica,* 58(1):71-93.

Raup, David M.

1991 *Extinction: Bad genes or bad luck?* (New York, W.W. Norton & Company Inc.) [A must-read before taking sides about reasons for extinction.]

Rogers, Everett M.

1995 *Diffusion of Innovations*. (New York, The Free Press) [A framework for understanding the transmission of change.]

Sanchez, Guadalupe

2004 *El Bajio: Recent Paleoindian Investigations in Sonora, Mexico.* Paper presented for the Society of American Archaeology.

Solberger, J.B.

1985 "A Technique for Folsom Fluting." *Lithic Technology,* 14(1): 41-50.

Speth, John D.

1972 "Mechanical Basis of Percussion Flaking." *American Antiquity,* 37(1):34-60.

Stanford, D.J.

1978 "The Jones-Miller Site: An Example of Hell Gap Bison Procurement Strategy." In *Bison Procurement and Utilization: A Symposium*, edited by L.B. Davis and M. Wilson, pp. 90-97. Society for American Archaeology.

2003 Personal communication.

Stanford, Dennis and Bruce Bradley

2000 "The Solutrean Solution." Scientific American Discovering Archaeology2(1):54-55.

2002 "Ocean Trails and Prairie Paths? Thoughts About Clovis Origins." In *The First Americans: The Pleistocene Colonization of the New World* pp. 255-271. Edited by N. G. Jablonski. Memoirs of the California Academy of Sciences No. 27. (Berkeley, University of California Press)

Stanford, Dennis and Jane Day, editors

1992 *Ice Age Hunters of the Rockies*. (Denver, University Press of Colorado)

Stanford, Dennis and Robert Patten

1984 "R-6, A preliminary Report of a Cody Site in North-Central New Mexico." *Papers of the Philmont Conference on the Archaeology of Northeastern New Mexico*. Edited by Carol J. Condie. 6(1):189-199.

Stoianivici, Dan and Yildirim Hurmuzlu

1996 "A Critical Study of the Applicability of Rigid Body Collision Theory." *ASME Journal of Applied Mechanics,* 63(2):307-316.

Straus, Lawrence Guy

1985 "Stone Age Prehistory of Northern Spain." *Science,* 230(4725):501-507.

Surovell, Todd A.

2004 Personal communication.

Tankersley, Kenneth B.

1998 "The Crook County Clovis Cache." *Current Research in the Pleistocene,* 15:86-88.

2002 *In Search of Ice Age Americans*. (Layton, Gibbs Smith Publishers)

Tunnell, Curtis, and LeRoy Johnson

1991 *Comparing Dimensions for Folsom Points and Their Byproducts from the Adair-Steadman and Lindenmeier Sites, As Well As a Few Other Localities*. Texas Historical Commission.

Tsirk, Are

1986 *Regarding Formation and Utility of a Class of Anomalous Wallner Lines on Obsidian*. Presented at Conference on the Fractography of Glasses and Ceramics, New York State College of Ceramics, Alfred University.

Weber, Bob

2001 Viewing of the Mockingbird Gap site artifacts.

William, Jerry D., Mathew J. Root, and Lisa K. Shifrin

1997 "Lake Ilo Ultrathin Bifaces and Folsom Points: Separate Production Sequences." *Current Research in the Pleistocene,* 14:111-112.

Wilmsen, Edwin N. and Frank H.H. Roberts, Jr.

1984 *Lindenmeier, 1934-1974: Concluding Report on Investigations*. (Washington, D.C., Smithsonian Institution Press) [The largest reported assemblage of Folsom artifacts.]

Glossary

angle of blow Angle between a flake's long axis and the vector direction of a blow.

anthropolithic Made out of stone by the hand of man .

anvil Rigid support for a percussion blow.

arris Raised intersection of flake scars.

BP Before Present. Actual years before 1960. *See radiocarbon years*

bevel Steep, wedge-shaped edge.

bulb of force Swelling thickness at the beginning of a flake.

cache Objects stored for provisional use at a later time. If associated with a burial, the use is projected to a spiritual plane.

center of mass The point where an object can be perfectly balanced.

chalcedony Fibrous quartz.

chert General term for silicate rock. Sometimes used to describe incompletely silicified flint.

channel flake Portion removed longitudinally from the base of a preform. *See channel scar, flute scar*

channel scar *See flute scar*

conchoidal fracture Rippled fracture in the shape of a clam shell.

context Relevance to a larger structure or background.

core A piece of stone intended to produce a flake to be used as a tool.

core tablet Special flake that rejuvenates a core platform by truncating its top.

complete flake Any flake that terminates at a free surface other than as a step.

damping Loss of energy by friction or fracture.

debitage French word for waste flakes. *See waste flakes*

dorsal Top surface, often concave. *See ventral*

dynamic loading Force responsible for flake detachment lasts less than one cycle of natural period of vibration. *See static loading.*

equifinality Condition where multiple means of reaching a result are not mutually exclusive.

facet A plane surface, used to characterize the geometry of an artifact.

face plane An imaginary, flat plane tangent to the face of a biface or core.

face planing Strategy or process control involving flattening a face by lateral or longitudinal flakes.

feathered flake Gently tapered flake termination, where the flake scar blends smoothly with the core face.

flute scar Long, convex scar left by removal of a channel flake. *See channel flake*

fluting The process of removing a channel flake from a preform face.

forensic Suitable for public forum.

form The general morphology of an object.

fractional flake Flake that fails to achieve theoretical maximum extent by some percentage. *See full-length flake*

full-length flake Flake that extends to the extreme edge from its origin. *See fractional flake*

gull wing Wing-shaped lines on a fracture surface that indicate interfering stress waves. *See Wallner wake*

Hertzian cone Fracture initiation associated with a small projectile impacting a surface at right angles.

impact site Place where blow is to land, whether prepared or not, need not be isolated. *See platform*

Levallois flake Middle Paleolithic strategy using an arris platform and specific surface isolation to control flake morphology.

natural frequency 1/(½ SQRT [M*K]) where M is mass and K is a constant.

obsidian Natural glass of volcanic origin.

outrepasse French for "over the edge." *See overshot flake*

overshot flake Flake that rounds off the opposite edge. *See outrepasse*

period of vibration The time required for an undamped system to complete one
cycle of free vibration.

platform Raised surface, implies isolated character. *See impact site.*

platform angle Angle between the impacted surface and the core face.

platform offset Distance from an impact site to a plane tangent to the core face. *See reduced margin*

preform Prior to final form (use where preliminary shaping has occurred.) *See blank, rough-out*

process step Portion of production distinguished by specific means of process control. *See stage, process-step control*

process-step control Means by which a process step is made reliable and predictable. *See process step, stage*

quarry blank Least refined stage of production as removed from the area of the quarry. *See rough-out*

radiocarbon years Uncorrected C-14 age. *See BP*

reduced margin Core or preform edge, deliberately modified to move the impact site further from the tangent to the face. *See platform offset*

rough-out Earliest stage of biface production, before product shape is established. *See quarry blank, preform, spall*

spall A fragment of stone from a quarry, before it has been altered.

stade Coldest point of an ice age cycle.

stage Episode of production activity that may span more than one process step. *See process step*

static loading Load is applied longer than one cycle of natural period of vibration. *See dynamic loading.*

strategy Rules and priorities that a knapper follows.

step flake Right-angled termination of a flake caused by loss of forward impulse.

technique Particular combination of tool and muscle used for a particular result.

technology Means used to escape the constraints of our environment.

tool inclination The low angle between the face of a core and the long axis of the hammer.

tradition Culturally specific order of business handed down over generations, including techniques and concepts.

typology Formal description of an artifact that can relate it to a specific time and culture.

ventral Bottom surface. *See dorsal*

Walner wake Arced lines on a fracture surface, formed by interfering stress waves. *See gull wing*

waste flakes Cast-off debris from flintknapping. *See debitage*

Index